ROBE

RESEARCH AND PERSPECTIVES IN ALZHEIMER'S DISEASE

Fondation Ipsen

Springer

Berlin
Heidelberg
New York
Barcelona
Budapest
Hong Kong
London
Milan
Paris
Santa Clara
Singapore
Tokyo

B.T. Hyman C. Duyckaerts
Y. Christen [Eds.]

Connections, Cognition and Alzheimer's Disease

With 63 Figures and 8 Tables

 Springer

Hyman, B.T., M.D., Ph.D.
Neurology Service
Massachusetts General Hospital
Fruit Street
Boston, MA 02114
USA

Duyckaerts, C., M.D., Ph.D.
Laboratoire de Neuropathologie R. Escourolle
Hôpital de La Salpêtrière, Blvd. de l'Hôpital
75651 Paris Cedex 16
France

Christen, Y., Ph.D.
Fondation IPSEN
24, rue Erlanger
75651 Paris Cedex 16
France

ISBN 3-540-62205-5 Springer-Verlag Berlin Heidelberg New York

Library of Congress Cataloging-in-Publication Data. Connections, cognition, and Alzheimer's disease / B.T. Hyman, C., Duyckaerts, Y. Christen (eds.). p. cm. – (Research and perspectives in Alzheimer's disease) Includes bibliographical references and index. ISBN 3-540-62205-5 (hardcover) 1. Alzheimer's disease-Pathophysiology. I. Hyman, B.T. II. Duyckaerts, C. III. Christen, Yves. IV. Series. [DNLM: 1. Alzheimer's Disease. 2. Synaptic Transmission. 3. Cognition Disorders-physiopathology. WT 155 C752 1997] RC523.C693 1997 618.97'6831-dc21 DNLM/DLC for Library of Congress

© Springer-Verlag Berlin Heidelberg 1997
Printed in Germany

Production: PRO EDIT GmbH, D-69126 Heidelberg

Cover design: Design & Production, D-69121 Heidelberg

Typesetting: Mitterweger Werksatz GmbH, Plankstadt

SPIN: 10560125   27/3136 – 5 4 3 2 1 0 – Printed on acid-free paper

Preface

Alzheimer's disease invades the brain from the inside. Unlike an abcess, a metastasis or an infarct, the disease follows specific tracks and avoids certain cortical areas while flourishing in others. Any observer is struck by the exquisite selectivity of the lesions and could, indeed, conclude that Alzheimer's disease knows neuroanatomy. However, should the term "disease" be used to define this disorder? Several genes, located on at least three different chromosomes, have been implicated in the disease. The ApoE4 genotype has been shown to be an important risk factor, but dementia pugilistica also suggests that environment can be involved in at least some aspects of the disorder. These data favor the now prevalent view that Alzheimer's disease should instead be considered as a syndrome, and probably all of the contributors to this volume are ready to endorse this point of view.

If "Alzheimer's syndrome" is the final common pathway to several pathogenetic mechanisms, there should be an event at one point in the course of the specific etiology that triggers a somewhat stereotypic diffusion process along some neural connections. Scientists who are fascinated by the way in which the nervous system has morphologically encoded its function after a long phylogenetic history are also fascinated by this pathological progression of Alzheimer's syndrome. These scientists can be easily identified – they have studied the anatomy of the brain, or at least have a deep interest in its they are directly involved in neuropsychology or keep it in mind; they are naturally inclined to believe that determining the phenotypic expression of Alzheimer's syndrome in the brain is not a trivial question. Finally they are the scientists who attended (or would have liked to have attended) the meeting held in Paris on May 20. 1996, under the auspices of the Ipsen Foundation. Will this book convey the enthusiasm and excitement of a full day of proposals, discussion, and suggestions? Will the reader grasp new ideas? Will he or she be ready to imagine new projects to better understand Alzheimer's syndrome and to more successfully fight its unbearable consequences? We hope the answer will be yes to each of these questions.

As will be immediately clear to the reader, this book deals with a cruel reality – real patients, real brains and real lesions. The studies described herein have been made possible by large networks of collaborators and interactions that an ever decreasing number of countries are able to provide, from wise lawmakers determining the legal conditions for medical inquiry and autopsy to experienced prosectors, ready to skillfully perfom a job, the sense of which might be wrongly interpreted by many. In this long chain of events, the patient is not forgotten and we would like to dedicate this volume to the numerous and anonymous people who have given a part of themselves for the better understanding and better treatment of Alzheimer's disease.

Charles Duyckaerts
Bradley Hyman
Yves Christen

Acknowledgments: The editors wish to thank Mary Lynn Gage for editorial assistance and Jacqueline Mervaillie for the organization of the meeting in Paris.

Contents

Contributors

Alford, M.
Department of Neurosciences, University of California at San Diego, La Jolla, CA 92093-0624, USA

Agid, Y.
Unité INSERM 289, Hôpital de La Salpêtrière, 47 Blvd. de l'Hôpital, 75651 Paris Cedex, France

Bennecib, M.
Laboratoire de Neuropathologie R. Escourolle, Hôpital de La Salpêtrière, 47 Blvd. de l'Hôpital, 75651 Paris Cedex 13, France

Braak, E.
Department of Anatomy, J. W. Goethe University, Theodor Stern Kai 7, 69590 Frankfurt, Germany

Braak, H.
Department of Anatomy, J. W. Goethe University, Theodor Stern Kai 7, 69590 Frankfurt, Germany

Buèe, L.
Unité INSERM 422, Place de Verdun, 59045 Lille Cedex, France

Buèe-Scherrer, V.
Unité INSERM 422, Place de Verdun, 59045 Lille Cedex, France

Callaham, L.
Department of Anatomy and Neurobiology, University Rochester, Medical Center, 601 Elmwood Avenue, Box 603, Rochester, NY 14642, USA

Clark, P.
Section on Functional Brain Imaging, NIMH Building 10, Room 4C110, 10 Center Drive, MSC 1366 Bethesda, MD 20892-1366, USA

Coleman, P. D.
Department of Anatomy and Neurobiology, University of Rochester Medical Center, 601 Elmwood Avenue, Box 603, Rochester, NY 14642, USA

Colle M.-A.
Laboratoire des Neuropathologie R. Escourolle Hôpital de La Salpêtrière, 47 Blvd. de l'Hôpital, 75651 Paris Cedex 13, France

Courtney, M.
Section on Functional Brain Imaging, NIMH Building 10, Room 4C110, 10 Center Drive, MSC 1366 Bethesda, MD 20892-1366, USA

Damasio, A. R.
Department of Neurology, University of Iowa, College of Medicine, Iowa City, IA 52242, USA

Damasio, H.
Department of Neurology, University of Iowa, College of Medicine, Iowa City, IA 52242, USA

David, J. P.
Unité INSERM 422, Place de Verdun, 59045 Lille Cedex, France

Delacourte, A.
Unité INSERM 422, Place de Verdun, 59045 Lille Cedex ,France

De Teresa, R.
Department of Neurosciences, University of California at San Diego, La Jolla, CA 92093-0624, USA

Duyckaerts C.
Laboratoire de Neuropathologie R. Escourolle, Hôpital de La Salpêtrière, 47 Blvd. de l'Hôpital, 75651 Paris Cedex 13, France

Flood, D. G.
Cephalon Inc., 145 Brandywine Pkwy, West Chester, PA 19380, USA

Games, D.
Athena Neurosciences, 800 Gateway Blvd., South San Francisco, CA 94080, USA

Gauvreau, D.
Projet Image, Hôpital Côte des Neiges, 4565, Chemin de la Reine Marie, Montréal, Quebec H3W IW5, Canada

Gomez-Isla, T.
Neurology Service, Massachusetts General Hospital, Fruit Street, Boston, MA 02114, USA

Grignon Y.
Laboratoire de Neuropathologie R. Escourolle, Hôpital de La Salpêtrière, 47 Blvd. de l'Hôpital, 75651 Paris Cedex 13, France

Hauw, J.-J.
Laboratoire de Neuropathologie R. Escourolle, Hôpital de La Salpêtrière, 47 Blvd. de l'Hôpital, 75651 Paris Cedex 13, France

Haxby, J. V.
Section on Functional Brain Imaging, NIMH Building 10, Room 4C110, 10 Center Drive, MSC 1366, Bethesda, MD 20892-1366, USA

Hoesen van, G. W.
Departments of Anatomy and Neurology, The University of Iowa, Iowa City, IA 52242, USA

Hof, P. R.
Fishberg Research Center for Neurobiology, Box 1065, Mt. Sinai Medical School, One Gustave Levy Place, NY 10029-6574, USA

Hyman, B. T.
Neurology Service, Massachusetts General Hospital, Fruit Street, Boston, MA 02114, USA

Iwai, A.
Department of Neurosciences, University of California at San Diego, La Jolla, CA 92093-0624, USA

Johnson-Wood, K.
Athena Neurosciences, 800 Gateway Blvd., South San Francisco, CA 94080, USA

Lee, M.
Athena Neurosciences, 800 Gateway Blvd., South San Francisco, CA 94080, USA

Lucassen, P. J.
Medical Pharmacology, LACDR, Sylvius Laboratory, University of Leiden, The Netherlands

Mallory, M.
Department of Neurosciences, University of California at San Diego, La Jolla, CA 92093-0624, USA

Masliah, E.
Department of Neurosciences and Pathology, University of California at San Diego, La Jolla, CA 92093-0624, USA

Morrison, J. H.
Fishberg Research Center for Neurobiology, Box 1065, Mt. Sinai School of Medicine, One Gustave Levy Place, NY 10029-6574, USA

Nes, J. A. P. van de
Department of Pathology, University of Nijmengen, The Netherlands

Nimchinsky, E. A.
Fishberg Research Center for Neurobiology, Box 1065, Mt. Sinai School of Medicine, One Gustave Levy Place, NY 10029-6574, USA

Olanow, C. W.
Department of Neurology, Box 1065, Mt. Sinai School of Medicine, One Gustave Levy Place, NY 10029-6574, USA

Petit, H.
Clinique Neurologique, Hôpital R. Salengro, 59037 Lille Cedex, France

Perl, D. P.
Fishberg Research Center for Neurobiology, Box 1065, Mt. Sinai Medical School, One Gustave Levy Place, NY 10029-6574, USA

Ravid, R.
Netherlands Brain Bank, Meibergdreef 33, 1105 AZ Amsterdam, The Netherlands

Robitaille, Y.
Projet Image, Hôpital Côte des Neiges, 4565 Chemin de la Reine Marie, Montréal, Quebec H3W IW5, Canada

Saitoh, T.
Department of Neurosciences, University of California at San Diego, La Jolla, CA 92093-0624, USA

Salehi, A.
Netherlands Institute for Brain Research, Meibergdreef 33, 1105 AZ Amsterdam The Netherlands and Department of Physiology, Faculty of Medicine, Shahid Beheshti University of Medical Sciences, Teheran, Iran

Schenk, D.
Athena Neurosciences, 800 Gateway Blvd., South San Francisco, CA 94080, USA

Sergeant, N.
Unité INSERM 422, Place de Verdun, 59045 Lille Cedex, France

Swaab, D. F.
The Netherlands Institute for Brain Research, Meibergdreef 33, 1105 AZ Amsterdam, Netherlands

Troncoso, C.
Laboratory for Neuropathology, John Hopkins University, 558 Ross Research Bldg., 720 Rutland Ave., Baltimore, MD 21205-2196, USA

Uchihara, T.
Laboratoire de Neuropathologie R. Escourolle, Hôpital de La Salpêtrière, 47 Blvd. de l'Hôpital, 75651 Paris Cedex 13, France

Ungerleider, L. G.
Laboratory of Psychology and Psychopathology, National Institute of Mental Health, Bethesda MD 20982, USA

Vermersch, P.
Clinique Neurologique, Hôpital R. Salengro, 59037 Lille Cedex, France and Unité INSERM 422, Place de Verdun, 59045 Lille Cedex, France

Wattez, A.
Unité INSERM 422, Place de Verdun, 59045 Lille Cedex, France

West, M. J.
Department of Neurobiology, Institute of Anatomy, University of Aarhus, 8000 Aarhus C, Denmark

Aspects of Cortical Destruction in Alzheimer's Disease

H. Braak[*], E. Braak

Summary

Alzheimer's disease is an immutably progressing dementing disorder. Its major pathologic hallmark is the development of cytoskeletal changes in a few susceptible neuronal types. These changes do not occur inevitably with advancing age, but once the disease has begun, spontaneous recovery or remissions are not observed.

The initial cortical changes develop in the poorly myelinated transentorhinal region of the medial temporal lobe. The destructive process then follows a predictable pattern as it extends into other cortical areas. Location of the tangle-bearing neurons and the severity of changes allow the distinction of six stages in disease propagation (transentorhinal stages I–II: clinically silent cases; limbic stages III–IV: incipient Alzheimer's disease; neocortical stages V–VI: fully developed Alzheimer's disease). A small number of cases display particularly early changes, indicating that advanced age is not a prerequisite for the evolution of the lesions. Alzheimer's disease is thus an age-related, but not an age-dependent, disease. The degree of brain destruction at stages III–IV often leads to the appearance of initial clinical symptoms, while stages V–VI represent fully developed Alzheimer's disease. Assessment of stage V–VI cases allows estimation of the rate of prevalence of the disease.

The pattern of appearance of the neurofibrillary changes bears a striking resemblance to the inverse sequence of cortical myelination. Factors released by oligodendrocytes exert important influence upon nerve cells and suppress disordered neuritic outgrowth. The lack of such factors due to premature dysfunction of oligodendrocytes could contribute to imbalances in the neuronal cytoskeleton and eventually initiate the development of neurofibrillary changes.

[*] Department of Anatomy, J. W. Goethe University, Theodor Stern Kai 7, D-60590 Frankfurt, Germany

B. T. Hyman / C. Duyckaerts / Y. Christen (Eds.)
Connections, Cognition, and Alzheimer's Disease
© Springer-Verlag Berlin Heidelberg 1997

Introduction

An early symptom of the brain destruction in Alzheimer's disease (AD) is a subtle decline of memory functions. The mild initial symptoms slowly worsen, and personality changes gradually appear, followed by deterioration of language functions, impairment in visuospatial tasks, and, in the end stages, motor dysfunction in the form of a hypokinetic hypertone syndrome (Reisberg et al. 1989, 1992; Berg and Morris 1994; Corey-Bloom et al. 1994). The steady aggravation of symptoms reflects the gradual expansion of brain destruction, which begins in a few limbic areas of the cerebral cortex, then spreads in a predictable, nonrandom manner across the hippocampus, the neocortex, and a number of subcortical nuclei (Kemper 1978; Hyman et al. 1984, 1990; Hyman and Gomez-Isla 1994; van Hoesen and Hyman 1990; van Hoesen et al. 1991; van Hoesen and Solodkin 1993; Arnold et al. 1991; Braak and Braak 1991, 1994; Price et al. 1991; Arriagada et al. 1992a, b).

Anatomical Considerations

Alzheimer's disease is largely a disorder of the cerebral cortex, the brain's chief controlling entity. For a better understanding of the typical pattern of AD-related cortical destruction, recall that the cortex consists of an extensive, more or less uniformly built neocortex and a small, heterogeneously composed allocortex (Brodmann 1910; Vogt and Vogt 1919; Braak 1980; Zilles 1990). The allocortex includes limbic system centers such as the hippocampal formation and the entorhinal region. The subcortical nuclear complex of the amygdala is closely related (Amaral and Insausti 1990; Amaral et al 1992). The parietal, occipital, and temporal neocortices are each comprised of a primary core field, a belt region, and related association areas. In the normal human brain, somato-sensory, visual, and auditory information proceeds through the respective neocortical sensory core and belt fields to a variety of association areas, from which the data are conveyed via long cortico-cortical pathways to the prefrontal association cortex. The remarkable extensiveness of the prefrontal cortex is a hallmark of the human brain. Tracts generated in this highest organizational level of the brain then transmit the data via the frontal belt to the primary motor area. The major routes of transport are the striatal loop and the cerebellar loop. Significant portions of the basal ganglia, many nuclei of the lower brain stem, and the cerebellum are incorporated in the regulation of cortical output via these loops (Fig. 1). Part of the stream of data from the neocortical sensory association areas to the prefrontal cortex branches off, eventually converging upon the entorhinal region and the amygdala (Felleman and Essen 1991; Amaral et al. 1992). Such connections comprise the afferent leg of the limbic loop. Neocortical information is thus the dominant source of input to the human limbic system. Further processing occurs in the entorhinal region, the amygdala, and the hippocampal formation. Projections from all these areas contribute of the efferent leg of the limbic loop, which exerts

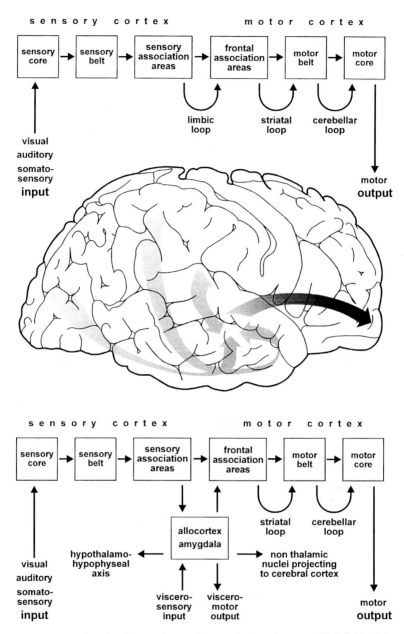

Fig. 1. Somato-sensory, visual, and auditory information proceeds through core and belt fields of the neocortex to a variety of association areas. The data are then transported via long cortico-cortical pathways to the prefrontal association cortex. From there, the data are transferred – preferably by way of the striatal and cerebellar loops – to the motor core field. Part of the stream of data from the sensory association areas to the prefrontal cortex branches off and converges upon the entorhinal region and the amygdala (afferent leg of the limbic loop). Projections from the entorhinal region, the amygdala, and the hippocampus exert important influence upon the prefrontal cortex (efferent leg of the limbic loop).

important influence on the prefrontal cortex (Alexander et al. 1990; Pandya and Yeterian 1990; Heimer et al. 1991; Braak et al. 1996). In the upper half of Figure 2, the limbic loop is shown in a bit more detail. The gray arrow emphasizes the strategic position of the limbic loop between the neocortical sensory association areas on the left and the prefrontal association cortex on the right. The hippocampal formation, the entorhinal region, and the amygdala are densely interconnected, and together the three represent the highest organizational level of the limbic system. All these limbic areas play a significant role in the maintenance of memory functions and in establishing the emotional aspects of personality (Amaral 1987; Squire and Zola-Morgan 1988, 1991; Hyman et al. 1990; Damasio and Damasio 1991; Zola-Morgan and Squire 1993).

Development of Alzheimer's Disease-Related Intraneuronal Pathology

The most outstanding neuropathologic feature of AD is a progressive deposition of abnormally phosphorylated and cross-linked tau proteins in a few susceptible types of nerve cells (Goedert et al. 1991; Arriagada et al. 1992a, b; Dickson et al. 1992; Goedert 1993; Iqbal et al. 1994; Price and Sisodia 1994; Vitek et al. 1994; Trojanowski et al. 1995). The intraneuronal material is generally deposited symmetrically in both hemispheres (Moossy et al. 1988; Arnold et al. 1991). Three forms in which the material occurs are neurofibrillary tangles (NFTs), neuropil threads (NTs), and argyrophilic components of neuritic plaques (NPs). NFTs develop within the nerve cell body, whereas NTs are found in the distal segments of dendrites (Braak et al. 1994). The first changes to occur are generally NTs and NFTs; NPs develop later (Braak and Braak 1991, 1994). Rather than appearing abruptly neurofibrillary changes develop gradually over a long period of time. Their structure changes remarkably as they appear, grow to maturity, and eventually disappear from the tissue. Only a few neuronal types are prone to

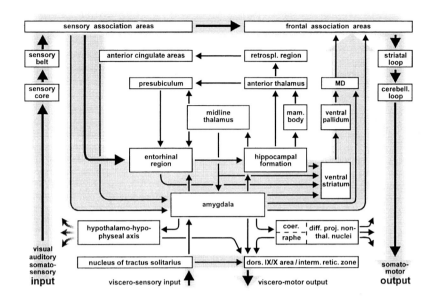

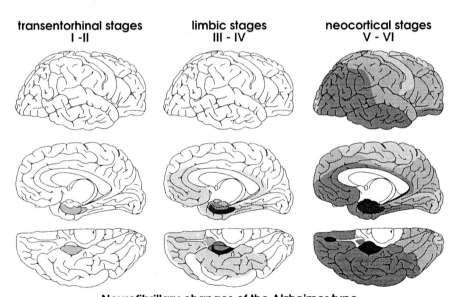

Neurofibrillary changes of the Alzheimer type

development of NFTs and NTs. Within the cerebral cortex, all NFT-bearing neurons are pyramidal (projection) cells. The entangled neurons eventually die. After deterioration of the parent cell, the pathologic material turns into an extraneuronal "ghost" tangle. In the course of this process, the NFT becomes less densely twisted and gradually loses its specific argyrophilia (Alzheimer 1907; Bancher et al. 1989; Probst et al. 1991; Braak et al. 1994).

Regional Distribution of Neurofibrillary Tangles and Neuropil threads

The spread of the intraneuronal changes is by no means uniform. NFTs and NTs, in particular, involve only specific architectonic units (Kemper 1978; Hyman et al. 1984, 1990; Hyman and Gomez-Isla 1994; van Hoesen and Hyman 1990; van Hoesen et al. 1991; van Hoesen and Solodkin 1993; Arnold et al. 1991; Braak and Braak 1991, 1994; Price et al. 1991). The transentorhinal region, located in the depths of the rhinal sulcus, usually shows the first lesions. Then, following a predictable pattern, the destruction extends into the hippocampus and the neocortex. The pattern of development of the lesions varies little among individuals and provides a basis for distinguishing six stages in the evolution of AD-related changes (lower half of Fig. 2 and Fig. 3).

Direct observation of the sequential development of the destructive process is impossible at present. The staging concept is thus necessarily derived from cross sectional data. In principal, the staging process is also artificial, since the AD-related changes spread continually rather than appearing in distinct steps. Categorization into six stages is recommended purely for practical reasons. A study of 2369 staged brains from non-selected autopsy cases – including non-demented and demented subjects aged 25 to 95 years – corroborates the assumption that early stages occur predominantly in relatively young individuals, whereas the more advanced stages gradually appear with increasing age (Ohm et al. 1995). Interestingly, a number of brains from prospective clinico-pathologic studies demonstrate a correlation between the results of the neuropathologic staging procedure and assessments of the intellectual status of these patients (Jellinger et al. 1991; Bancher et al. 1993; Braak et al. 1993; Duyckaerts et al. 1994, 1995).

Transentorhinal Stages I and II

Only the transentorhinal region is consistently involved in the most mildly affected cases. It probably serves as the main port of entrance for neocortical data being transmitted to limbic components of the medial temporal lobe (Figs. 2, 3). Specific projection cells in this region are the first neurons in the cerebral cortex to show the development of NFTs and NTs. In stage I, only a few transentorhinal projection cells are changed, whereas in stage II numerous tangle-bearing cells are found with others in the entorhinal region proper. Stage II

Neurofibrillary changes of the Alzheimer type

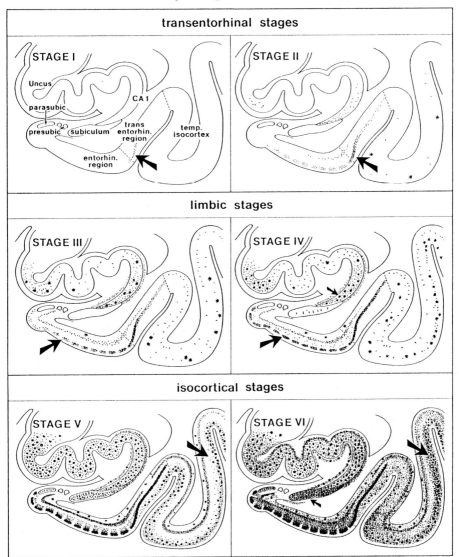

Fig. 3. Summary diagram of neurofibrillary changes seen in the hippocampal formation, entorhinal region, and adjoining temporal neocortex. Note the development of changes from stage I to stage VI. The arrows designate the key features. CA1, first sector of the Ammon's horn; entorhin., entorhinal; parasub, parasubiculum; presubic, presubiculum; temp., temporal; transentorhin., transentorhinal. (Reproduced with permission from Braak and Braak 1991.)

Fig. 4. Development of neurofibrillary changes in a total number of 2369 non-selected autopsy ▶ cases. The first line displays the percentage of cases devoid of neurofibrillary changes to the total number of cases within each age category. The brains of some old-aged individuals are free of neurofibrillary changes. The 2nd, 3rd, and 4th lines are similarly designed, and show the various degrees in the evolution of the AD-related intraneuronal changes. The pattern seen in these illustrations is stable and is virtually unchanged with addition of further cases. Some individuals develop initial stage I/II (transentorhinal) pathology early in life. Thus, old age is not an indispensable prerequisite for the evolution of the changes. Since spontaneous remissions are not observed, stage I lesions mark the beginning of AD. The third line represents cases of the limbic stages in which initial symptoms are frequently displayed, while end stage (= neocortical) cases with fully developed AD are shown in the fourth line. In spite of the differences between epidemiologic and postmortem studies, there is a fairly close correspondence between clinically evaluated prevalence rates of AD and the results displayed in line four. Note that early stages occur preferably in younger age categories, while more advanced stages gradually appear with increasing age.

pathology slightly impedes the transmission of neocortical information – via entorhinal region – to the hippocampal formation.

As in many other neurodegenerative diseases, the pathologic lesions develop over a period of years at their predilection sites without exceeding the threshold that produces clinical symptoms. The transentorhinal stages I and II present with no obvious impairments of intellectual capacities and are thus considered to represent the preclinical phase of AD.

Neurofibrillary changes of the Alzheimer type do not inevitably accompany old age, and there is no continuum uniting the specific age-related changes in the human brain and the AD-related intraneuronal pathology (Dickson et al. 1991). The first line of Figure 4 displays the percentage of cases with complete absence of neurofibrillary changes for various age categories in a sample of 2369 non-selected autopsy cases. Some individuals show no intraneuronal changes et all, even in old age.

Intraneuronal NFTs/NTs and a variable number of extraneuronal "ghost" tangles are found in all stage I–VI cases. Just "ghost" tangles alone were never found. "Fresh" NFTs and NTs were present in all cases, showing that the pathologic process was still going on at the time of death. This finding indicates that spontaneous remission does not occur in AD. Once the destructive process has begun, it progresses relentlessly. The first NFTs/NTs are very significant, as they mark the beginning of the disease.

A small proportion of cases develop initial NFTs/NTs at a surprisingly young age. Obviously, advanced age is no prerequisite for the development of the intraneuronal pathology. This observation casts doubt on theories that seek to explain the changes as a consequence of noxious influences that are generally expected to take effect in old age (peroxidative stress, mitochondrial dysfunction, or imbalance of glucose metabolism). This does not rule out the possibility that peroxidative stress may contribute to the changes in advanced stages of the disease or influence the pace of the pathologic process (Volicer and Crino 1990; Pappolla

Development of neurofibrillary changes (n=2369)

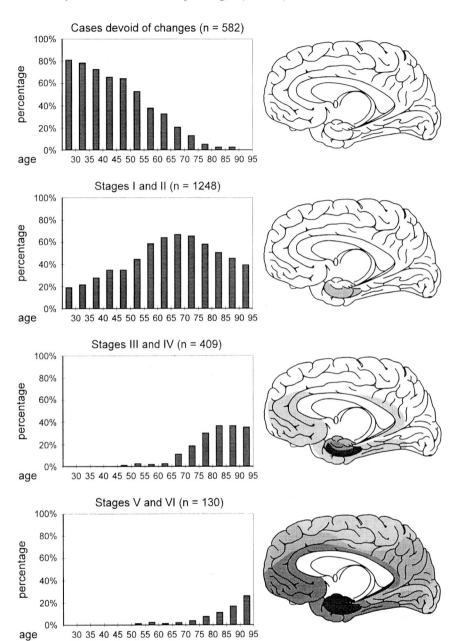

et al. 1992; Balazs and Leon 1994; Benzi and Moretti 1995; Choi 1995). However, it is probably not a primary factor in the pathogenesis of the initial AD-related lesions.

Upon postmortem evaluation, stage I cases are easily distinguishable from normal brains free of neurofibrillary changes. As mentioned already, stage I cases are clinically silent. The real rate of incidence of AD cannot be evaluated in a living population because the initial phase remains asymptomatic. However, data regarding the appearance of new cases have particular significance in providing information about the spread of the disease. Postmortem studies using a staging procedure provide such data as summarized in the second line of Figure 4.

Limbic Stages III and IV

In cases at the limbic stages III and IV, cortical destruction is already severe, but is limited to a few allocortical regions and adjoining areas (Figs. 2, 3). Since neocortical destruction is not present or mild, stages III and IV fail to meet commonly used criteria for neuropathologic diagnosis of AD (Khachaturian 1985; Tierney et al. 1988; Mirra et al. 1991; Gearing et al. 1995). The key feature of the limbic stages is the striking destruction of the specific entorhinal cellular layers responsible for the data transfer from the neocortex to the hippocampal formation and vice versa (Figs. 1–3). Initially, the hippocampal formation itself is only mildly involved; at stage IV, however, the destructive process spreads out markedly from the entorhinal region into the amygdala, the hippocampal formation, and, in particular, into the adjoining association areas of the temporal neocortex.

The clinical protocols of many individuals at stage III or IV record an impairment of cognitive functions and the presence of subtle personality changes. In others, the appearance of symptoms is still camouflaged by individual reserve capacities that compensate for the local destruction. At first glance, it is surprising that the destruction of just two entorhinal cellular layers could cause functional disturbances. This is explained by the fact that the specific lesions bilaterally hamper both the feed forward and the feedback projections necessary for normal function of the neocortex-hippocampus interrelation. The local pathologic process interrupts the limbic loop at multiple sites. Because of the tendency toward expression of initial clinical symptoms and the characteristic brain lesions, stage III or IV cases are considered to represent incipient AD.

Neocortical Stages V and VI

The final stages of AD show large numbers of NFTs and NTs in virtually every subdivision of the cerebral cortex. A key feature of stage V is the extremely severe destruction of neocortical association areas, leaving only the primary motor field, the primary sensory areas, and their belt regions uninvolved or only mildly affected. At stage VI, the pathologic process extends into the primary areas

(Figs. 2, 3). Stages V and VI correspond to the conventionally used criteria for neuropathologic confirmation of the clinical diagnosis of AD. In spite of the differences between epidemiologic and postmortem studies, there is a relatively high degree of correspondence between the neuropathologic data (fourth line in Fig. 4) and the clinically evaluated rates of prevalence of AD (Mortimer 1988; Hofmann et al. 1991; Bachman et al. 1993; Harrell et al. 1993; Ebly et al. 1994; Katzman and Kawas 1994; Corrada et al. 1995).

Progression of AD-Related Neurofibrillary Changes Repeats the Progression of Cortical Myelination in Reverse Order

The striking consistency of the lesional pattern that gradually develops in the course of Alzheimer's disease is still an enigma. Figures 5 and 6 illustrate that the sequence and pattern of AD-related destruction are strikingly similar to the progress of cortical myelination in early development, but in reverse order. Poorly myelinated cortical areas develop NFTs/NTs earlier in the course of AD and at higher density than do densely myelinated areas (Braak and Braak 1996).

Myelin appears and matures in an ordered sequence during ontogenesis. The human cortex exhibits particularly late and prolonged myelination, lasting well into adhulthood (Flechsig 1920; Yakovlev and Lecours 1967). The first traces of myelin appear in the neocortical core areas. The process then spreads through the border fields into the related association areas. The result is a remarkably dense myelination of the neocortical core fields in the brain of the human adult. A gradual decrease in the average density of myelin is observed with increasing distance from the core fields. The temporal association areas close to the transentorhinal region and the transentorhinal region itself are particularly sparsely myelinated (Vogt and Vogt 1919). Precisely this region is the site of the first neurofibrillary changes. The process extends gradually into the neocortex, successively reaching association areas of the third and the second temporal gyri. Involvement of the auditory core field and other core areas is observed only in the end stages of the disease (Figs. 4, 5).

Regressive changes in components of the brain often repeat the process of their maturation, though in reverse order. The similarities between cortical myelination and development of AD-related changes can be explained by postulating premature dysfunction of cortical oligodendrocytes. Factors released by oligodendrocytes influence neighboring neurons and suppress disordered neuritic outgrowth (Vaughan 1984; Schwab 1990; Cadelli et al. 1992; Hardy and Reynolds 1993; Kapfhammer and Schwab 1994). Stability of nerve cells increases with myelination of their axons, and age-related instability may thus be the result of decreasing influence of oligodendrocytes. One speculative scenario is that a lack of oligodendrocytic factors results in alterations of the neuronal cytoskeleton and eventually in the induction of NFTs/NTs (Braak and Braak 1996).

Progression of cortical myelination

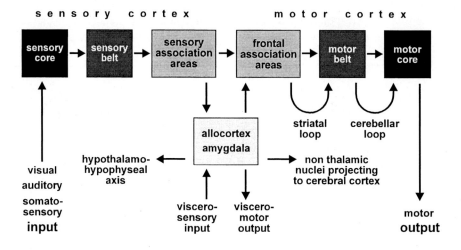

Progression of Alzheimer's disease related destruction

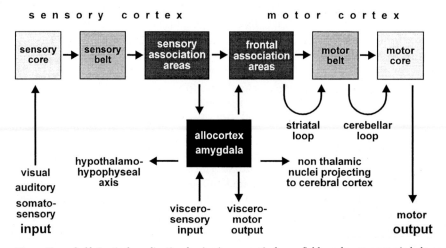

Fig. 5. *Upper half:* Cortical myelination begins in neocortical core fields and progresses via belt areas to the related association areas (indicated by darker shading). *Lower half:* AD-related cortical destruction begins in the transentorhinal region, from which the changes extend into adjoining areas, eventually reaching the neocortical core fields. Note that the sequence of destruction inversely recapitulates that of cortical myelination. (Reproduced with permission from Braak and Braak 1996.)

Progression of cortical myelination

Progression of Alzheimer's disease related destruction

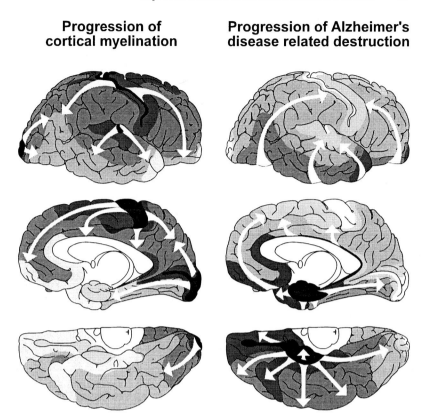

Fig. 6. Drawings of a right hemisphere showing the outward progression of myelination from the neocortical core fields into the association areas on the left side, and the progression of AD-related neurofibrillary changes from the transentorhinal and entorhinal regions via neocortical association fields and belt areas into the core fields on the right. Note the inverse pattern of the two processes. (Reproduced with permission from Braak and Braak 1996).

Acknowledgments

This study was kindly supported by grants from the Deutsche Forschungsgemeinschaft, the Bundesministerium für Forschung und Technologie, and Degussa, Hanau. The skillful assistance of Ms Szasz (drawings) is gratefully acknowledged.

References

Alexander GE, Crutcher MD, DeLong MR (1990) Basal ganglia-thalamocortical circuits: parallel substrates for motor, oculomotor, "prefrontal" and "limbic" functions. Progr Brain Res 85: 119–146
Alzheimer A (1990) Über eine eigenartige Erkrankung der Hirnrinde. Centralbl Nervenheilk Psychiatr (Leipzig) 30: 177–179

Amaral DG (1987) Memory: anatomical organization of candidate brain regions. In: Brookhart JM, Mountcastle VB (eds) Handbook of physiology: The nervous system, V: Higher functions of the nervous system. 5th ed. Amer Physiol Soc, Bethesda, pp 211–294

Amaral DG, Insausti R (1990) Hippocampal formation. In: Paxinos E (ed) The human nervous system. Academic Press, New York, pp 711–756

Amaral DG, Price JL, Pitkänen A, Carmichael ST (1992) Anatomical organization of the primate amygdaloid complex. In: Aggleton JP (ed) The amygdala: Neurobiological aspects of emotion, memory, and mental dysfunction. Wiley-Liss, New York, pp 1–66

Arnold SE, Hyman BT, Flory J, Damasio AR, van Hoesen GW (1991) The topographical and neuroanatomical distribution of neurofibrillary tangles and neuritic plaques in the cerebral cortex of patients with Alzheimer's disease. Cerebral Cortex 1: 103–116

Arriagada PV, Growdon JH, Hedley-Whyte ET, Hyman BT (1992a) Neurofibrillary tangles but not senile plaques parallel duration and severity of Alzheimer's disease. Neurology 42: 631–639

Arriagada PV, Marzloff L, Hyman BT (1992b) Distribution of Alzheimer-type pathologic changes in non-demented elderly individuals matches the pattern in Alzheimer's disease. Neurology 42: 1681–1688

Bachman DL, Wolf PA, Linn RT, Knoefel JE, Cobb JL, Belanger AJ, White LR, D'Agostino RB (1993) Incidence of dementia and probable Alzheimer's disease in a general population: the Framingham study. Neurology 43: 515–519

Balazs L, Leon M (1994) Evidence of an oxidative challenge in the Alzheimer's brain. Neurochem Res 19: 1131–1137

Bancher C, Brunner C, Lassmann H, Budka H, Jellinger K, Wiche G, Seitelberger F, Grundke-Iqbal I, Wisniewski HM (1989) Accumulation of abnormally phosphorylated τ precedes the formation of neurofibrillary tangles in Alzheimer's disease. Brain Res 477: 90–99

Bancher C, Braak H, Fischer P, Jellinger KA (1993) Neuropathological staging of Alzheimer lesions and intellectual status in Alzheimer's and Parkinson's disease. Neurosci Lett 162: 179–182

Benzi G, Moretti A (1995) Are reactive oxygen species involved in Alzheimer's disease? Neurobiol Aging 16: 661–674

Berg L, Morris JC (1994) Diagnosis. In: Terry RD, Katzman R, Bick KL (eds) Alzheimer Disease. Raven Press, New York, pp 9–25

Braak E, Braak H, Mandelkow EM (1994) A sequence of cytoskeleton changes related to the formation of neurofibrillary tangles and neuropil threads. Acta Neuropathol 87: 554–567

Braak H (1980) Architectonics of the human telencephalic cortex. Springer, Berlin.

Braak H, Braak E (1991) Neuropathological stageing of Alzheimer-related changes. Acta Neuropathol. 82: 239–259

Braak H, Braak E (1994) Pathology of Alzheimer's disease. In: Calne DB (ed) Neurodegenerative diseases. Saunders, Philadelphia, pp 585–613

Braak H, Braak E (1996) Development of Alzheimer-related neurofibrillary changes in the neocortex inversely recapitulates cortical myelogenesis. Acta Neuropathol. 92: 197–201

Braak H, Duyckaerts C, Braak E, Piette F (1993) Neuropathological staging of Alzheimer-related changes correlates with psychometrically assessed intellectual status. In: Corian B, Iqbal K, Nicolini M, Winblad B, Wisniewski H, Zatta P (eds) Alzheimer's disease: Advances in Clinical and Basic Research. Third International Conference on Alzheimer's Disease and Related Disorders. Wiley, Chichester, pp 131–137

Braak H, Braak E, Yilmazer D, de Vos RAI, Jansen ENH, Bohl J (1996) Pattern of brain destruction in Parkinson's and Alzheimer's diseases. J Neural Transm 103: 455–490

Brodmann K (1910) Feinere Anatomie des Großhirns. In: Lewandowsky M (ed) Handbuch der Neurologie, Vol. 1, Springer, Berlin, pp 206–307

Cadelli DS, Bandtlow CE, Schwab ME (1992) Oligodendrocyte- and myelin-associated inhibitors of neurite outgrowth: their involvement in the lack of CNS regeneration. Exp Neurol 115: 189–192

Choi BH (1995) Oxidative stress and Alzheimer's disease. Neurobiol Aging 16: 675–678

Corey-Bloom J, Galasko D, Thal LJ (1994) Clinical features and natural history of Alzheimer's disease. In: Calne DP (ed) Neurodegenerative diseases. Saunders, Philadelphia, pp 631–645

Corrada M, Brookmeyer R, Kawas C (1995) Sources of variability in prevalence rates of Alzheimer's disease. Int J Epidemiol 24: 1000–1005

Damasio AR, Damasio H (1991) Disorders of higher brain function. In: Rosenberg RN (ed) Comprehensive neurology. Raven Press, New York, pp 639–657

Dickson DW, Crystal HA, Mattiace LA, Masur DM, Blau AD, Davies P, Yen SH, Aronson MK (1991) Identification of normal and pathological aging in prospectively studied nondemented elderly humans. Neurobiol Aging 13: 179–189

Dickson DW, Ksiezak-Reding H, Liu WK, Davies P, Crowe A, Yen S HC (1992) Immunocytochemistry of neurofibrillary tangles with antibodies to subregions of tau protein: identification of hidden and cleaved tau epitopes and a new phosphorylation site. Acta Neuropathol 84: 596–605

Duyckaerts C, He Y, Seilhean D, Delaère P, Piette F, Braak H, Hauw JJ (1994) Diagnosis and staging of Alzheimer's disease in a prospective study involving aged individuals. Neurobiol Aging 15 (Suppl 1): 140–141

Duyckaerts C, Delaère P, He Y, Camilleri S, Braak H, Piette F, Hauw JJ (1995) The relative merits of tau- and amyloid markers in the neuropathology of Alzheimer's disease. In: Bergener M, Finkel SI (eds) Treating Alzheimer's and other dementias. Springer, New York, pp 81–89

Ebly EM, Parhad IM, Hogan DB, Fung TS (1994) Prevalence and types of dementia in the very old: results from the Canadian study of health and aging. Neurology 44; 1593–1600

Felleman DJ, van Essen DC (1991) Distributed hierarchical processing in the primate cerebral cortex. Cerebral Cortex 1: 1–47

Flechsig P (1920) Anatomie des menschlichen Gehirns und Rückenmarks auf myelogenetischer Grundlage. Thieme, Leipzig

Gearing M, Mirra SS, Hedreen JC, Sumi SM, Hansen LA, Heyman A (1995) The consortium to establish a registry for Alzheimer's disease (CERAD). Part X. Neuropathology confirmation of the clinical diagnosis of Alzheimer's disease. Neurology 45: 461–466

Goedert M (1993) Tau protein and the neurofibrillary pathology of Alzheimer's disease. Trends Neurosci 16: 460–465

Goedert M, Spillantini MG, Crowther RA (1991) Tau proteins and neurofibrillary degeneration. Brain Pathol 1: 279–286

Hardy R, Reynolds R (1993) Neuron-oligodendroglial interactions during central nervous system development. J Neurosci Res 36: 121–126

Harrell LE, Callaway R, Powers R (1993) Autopsy in dementing illness: who participates? Alzheimer Dis Assoc Disord 7: 80–87

Heimer L, de Olmos J, Alheid GF, Zaborszky L (1991) "Perestroika" in the basal forebrain: opening the border between neurology and psychiatry. Progr Brain Res 87: 109–165

Hofman A, Rocca W, Brayne C (1991) The prevalence of dementia in Europe: a collaborative study of the 1980–1990 findings. Int J Epidemiol 20: 736–748

Hyman BT, van Hoesen GW, Damasio AR, Barnes CL (1984) Alzheimer's disease: cell-specific pathology isolates the hippocampal formation. Science 225: 1168–1170

Hyman BT, van Hoesen GW, Damasio AR (1990) Memory-related neural systems in Alzheimer's disease: an anatomic study. Neurology 40: 1721–1730

Hyman BT, Gomez-Isla T (1994) Alzheimer's disease is a laminar, regional, and neural system specific disease, not a global brain disease. Neurobiol Aging 15: 353–354

Iqbal K, Alonso AC, Gong CX, Khatoon S, Singh TJ, Grundke-Iqbal I (1994) Mechanism of neurofibrillary degeneration in Alzheimer's disease. Molecular Neurobiol 9: 119–123

Jellinger K, Braak H, Braak E, Fischer P (1991) Alzheimer lesions in the entorhinal region and isocortex in Parkinson's and Alzheimer's diseases. Ann New York Acad Sci 640: 203–209

Kapfhammer JP, Schwab ME (1994) Inverse patterns of myelination and GAP-43 expression in the adult CNS: Neurite growth inhibitors as regulators of neuronal plasticity. J Comp Neurol 340: 194–206

Katzman R, Kawas C (1994) The epidemiology of dementia and Alzheimer's disease. In: Terry RD, Katzman R, Bick KL (eds) Alzheimer's disease. Raven Press, New York, pp 105–122

Kemper TL (1978) Senile dementia: a focal disease in the temporal lobe. In: Nandy E (ed) Senile dementia: a biomedical approach. Elsevier, Amsterdam, pp 105–113

Khachaturian ZS (1985) Diagnosis of Alzheimer's disease. Arch Neurol 42: 1097–1105

Mirra SS, Heyman A, McKeel D, Sumi SM, Crain BJ, Brownlee LM, Vogel FS, Hughes JP, van Belle G, Berg L (1991) The consortium to establish a registry for Alzheimer's disease (CERAD). II. Standardization of the neuropathologic assessment of Alzheimer's disease. Neurology 41: 479–486

Moossy J, Zubenko GS, Martinez AJ, Rao GR (1988) Bilateral symmetry of morphological lesions in Alzheimer's disease. Arch Neurol 45: 251–254

Mortimer J (1988) Epidemiology of dementia: international comparisons. In: Brody JA, Maddox GL (eds) Epidemiology and aging: an international perspective. Springer, New York, pp 150–164

Ohm TG, Müller H, Braak H, Bohl J (1995) Close-meshed prevalence rates of different stages as a tool to uncover the rate of Alzheimer's disease-related neurofibrillary changes. Neuroscience 64: 209–217

Pandya DN, Yeterian EH (1990) Prefrontal cortex in relation to other cortical areas in rhesus monkey: architecture and connections. Progr Brain Res 85: 63–94

Pappolla MA, Omar RA, Kim KS, Robakis NK (1992) Immunohistochemical evidence of antioxidant stress in Alzheimer's disease. Am J Pathol 140: 621–628

Price DL, Sisodia SS (1994) Cellular and molecular biology of Alzheimer's disease and animal models. Ann Rev Med 45: 435–446

Price JL, Davis PB, Morris JC, White DL (1991) The distribution of tangles, plaques and related immunohistochemical markers in healthy aging and Alzheimer's disease. Neurobiol Aging 12: 295–312

Probst A, Langui D, Ulrich J (1991) Alzheimer's disease: a description of the structural lesions. Brain Pathol 1: 229–239

Reisberg B, Ferris SH, Kluger A, Franssen E, DeLeon MJ, Mittelman M, Borenstein J, Rameshwar K, Alba R (1989) Symptomatic changes in CNS aging and dementia of the Alzheimer type: Cross-sectional, temporal, and remediable concomitants. In: Bergener M, Reisberg B (eds) Diagnosis and treatment of senile dementia. Springer, Berlin, pp 193–223

Reisberg B, Pattschull-Furlan A, Franssen E, Sclan SG, Kluger A, Dingcong L, Ferris SH (1992) Dementia of the Alzheimer type recapitulates ontogeny inversely on specific ordinal and temporal parameters. In: Kostovic I, Knezevic S, Wisniewski HM, Spillich GJ (eds) Neurodevelopment, aging and cognition. Birkhäuser, Boston, pp 345–369

Schwab ME (1990) Myelin-associated inhibitors of neurite growth and regeneration in the CNS. Trends Neurosci 13: 452–456

Squire LR, Zola-Morgan S (1988) Memory: brain systems and behavior. Trends Neurosci 11: 170–175

Squire LR, Zola-Morgan S (1991) The medial temporal lobe memory system. Science 253: 1380–1386

Tierney MC, Fisher RH, Lewis AJ, Zorzitto ML, Snow WG, Reid DW, Nieuwstraten P (1988) The NINCDS-ADRDA work group criteria for the clinical diagnosis of probable Alzheimer's disease: a clinicopathologic study of 57 cases. Neurology 38: 359–364

Trojanowski JQ, Shin RW, Schmidt ML, Lee VMY (1995) Relationship between plaques, tangles, and dystrophic processes in Alzheimer's disease. Neurobiol Aging 16: 335–340

van Hoesen GW, Hyman BT (1990) Hippocampal formation: anatomy and the patterns of pathology in Alzheimer's disease. Progr Brain Res 83: 445–457

van Hoesen GW, Solodkin A (1993) Some modular features of temporal cortex in humans as revealed by pathological changes in Alzheimer's disease. Cerebral Cortex 3: 465–475

van Hoesen GW, Hyman BT, Damasio AR (1991) Entorhinal cortex pathology in Alzheimer's disease. Hippocampus 1: 1–8

Vaughan DW (1984) The structure of neuroglial cells. In: Jones EG, Peters A (eds) Cerebral cortex, vol. 2: Functional properties of cortical cells. Plenum Press, New York, pp 285–329

Vitek MP, Bhattacharya K, Glendening JM, Stopa E, Vlassara H, Bucala R, Monogue K, Cerami A (1994) Advanced glycation end products contribute to amyloidosis in Alzheimer's disease. Proc Natl Acad Sci USA 91: 4766–4770

Volicer L, Crino PB (1990) Involvement of free radicals in dementia of the Alzheimer's type: a hypothesis. Neurobiol Aging 11: 567–571

Vogt C, Vogt O (1919) Allgemeinere Ergebnisse unserer Hirnforschung. J Psychol Neurol 25: 279–461

Yakovlev PI, Lecours AR (1967) The myelogenetic cycles of regional maturation of the brain. In: Minkowski A (ed) Regional development of the brain in early life. Blackwell, Oxford, pp 3–70

Zilles K (1990) Cortex: In: Paxinos G (ed) The human nervous system. Academic Press, New York, pp 757–802

Zola-Morgan S, Squire LR (1993) Neuroanatomy of memory. Ann Rev Neurosci 16: 547–563

Cortical Feedforward and Cortical Feedback Systems in Alzheimer's Disease

*G. W. Van Hoesen**

Summary

This chapter provides a brief review of the principles of organization that relate to long association axons systems in the cerebral cortex. In particular, feedforward sensory systems are traced to their endstations in the entorhinal/hippocampal cortex, the amygdala and the nucleus basalis of Meynert. Feedback projections to the cerebral cortex from these limbic structures are also highlighted. These cortical connections are discussed relative to the topography of pathology in Alzheimer's disease and the fact that neurofibrillary tangles invariantly affect these cortical systems early in the illness. It is argued that the co-occurrence of pathology in the endstations of feedforward systems and the origin of initial feedback systems is coupled tightly to alterations of memory in Alzheimer's disease and other cognitive changes associated with the disorder. Widespread association cortex pathology, seen at endstage Alzheimer's disease, is related likely to degree, or density, of impairment in the disorder, but may be secondary to the behaviorally disruptive consequences of early and invariant limbic system pathology.

Introduction

Corticocortical connections have been studied intensively using experimental neuroanatomical methods for the past two and one-half decades and many organizational principles have emerged that characterize them (Pandya and Yeterian 1985; Goldman-Rakic 1988). The axons that form these connections fall into many categories. For example, some axons are unusually long and link the various lobes of the cerebral hemisphere together. These are the classic association systems of early dissections or the intrahemispheric axons. Others are shorter and link adjacent gyri or cytoarchitectural areas. These axons are often called U fibers. A large category of axons stay within the cortical gray matter and link its various layers and columns together. Some of the latter arise from interneurons, or local circuit neurons, characterized chemically for various peptides and en-

---

* Departments of Anatomy and Neurology, The University of Iowa, Iowa City, IA 52242 USA

B. T. Hyman / C. Duyckaerts / Y. Christen (Eds.)
Connections, Cognition, and Alzheimer's Disease
© Springer-Verlag Berlin Heidelberg 1997

zymes (Jones 1990). In general, long cortical axons always have at least some local collaterals before they enter the white matter, and local circuit neuron axons bifurcate extensively. The terminal endings of long association axons are generally conservative in the sense that they collateralize mainly in the vicinity of their targets and seldom give rise to corticofugal collaterals that leave the cortex.

The intent of this brief review is to summarize the major principles of cortical association axons that relate to their origin and termination and to comment on the feedforward and feedback streams of axons that characterize cortical connectivity. Special attention will be paid to the limbic cortices and subcortical limbic structures that are primary targets for feedforward cortical axons as well as the origin for feedback cortical axons. The disruption of these in Alzheimer's disease will be highlighted.

Cortical Feedforward Axons

A major strategy to understand the elaboration of sensory stimuli in the cortex for cognition has entailed focusing on the primary sensory cortices and assessing the dissemination of their axonal projections to other cortical areas. Inherent in this approach is the assumption that the sensory cortex is the first registry, receiving, so to speak, the raw material that peripheral receptors are exposed to. After a series of sequential projections to successively more distal cortical association areas, a more refined and accurate percept is thought to be generated (Gross 1992; Van Hoesen 1993; Zeki 1993). This hierarchical logic has deep roots in the clinicoanatomical study of cortical stroke (Gerschwind 1965), and experimental neuroanatomical findings are consistent with it (Pandya and Kuypers 1969; Jones and Powell 1970). The term "feedforward connections" has been coined to embody both the anatomy and the functional rationale for viewing cortical axons in this manner.

In the visual cortices, for example, the primary visual cortex or V_1 (Brodmann's area 17) sends axons to so-called early visual association areas in its vicinity. The cells of origin for these axons are located in layer III and the axons end in and around layer IV of the visual association cortex. The latter give rise to axonal projections to other more distally located visual association areas with the same pattern for cell of origin and axon termination (Rockland and Pandya 1979, 1981). Similar feedforward systems have been described for the primary auditory and somatosensory cortices (Jones and Powell 1970; Galaburda and Pandya 1983).

The targets of feedforward cortical axon projections are similar for all modalities. One is the multimodal association cortex of the frontal lobe and the multimodal association cortex of the superior temporal sulcus and posterior parahippocampal gyrus of the temporal lobe. These areas are designated multimodal since they receive cortical axons from association cortex of two or more modalities (Pandya and Yeterian 1985). Another target is the entorhinal cortex of the anterior parahippocampal gyrus which receives both sensory-specific association cortex axons and axons from multimodal association areas (Van Hoesen 1982). A third and equally sizable target is basal forebrain structures like the amygdala (Herzog and Van Hoesen 1976) and nucleus basalis of Meynert (Mesulam and Mufson 1984). It could be argued that the entorhinal cortex and the basal forebrain parts of the limbic system (amygdala and substantia innominata, including the nucleus basalis of Meynert) are the true endstations of cortical feedforward projections since each receive frontal and temporal multimodal association cortex input as well.

The above is a sketch or skeleton of feedforward cortical axons which addresses primarily the net outflow for a given sensory system. In fact, there are many discrete elements. In the visual cortex, for example, many ships are launched to association area destinations that have unique functions, and individually, these give rise to unique streams of axons that course to more distal association areas (Zeki 1975, 1990; Zeki and Shipp 1988). Cross-links between some of them diminish enthusiasm for strict sequential processing. Thus, the feedforward axons of the cortical visual system are best viewed as distributed hierarchical parallel systems with many unique functional properties (Van Essen et al. 1990; Felleman and Van Essen 1991). These ultimately end and converge on the multimodal association areas and the limbic system structures discussed above. Overall, layer III is the major source of feedforward axons, but as the endstations are approached, layer V makes a contribution as well. In the early association areas layer IV is the primary target of axons, but distally, columnar patterns of termination are not unusual (Barbas 1986).

Cortical Feedback Axons

Feedforward axonal projections are reciprocated by feedback axons. This is observed more clearly in the early association areas, where return projections course back to the primary sensory area from which they received projections. Unlike feedforward axons, which arise largely from layer III and to a lesser extent layer V, feedback axons arise from layer VI. Also, unlike feedforward axons that end in layer IV, feedback axons end most heavily in layer I (Rockland and Pandya 1979, 1981). This system of cortical axons is less understood than the feedforward system, but given what is known, it seems less bound by rules. For example, some distal association areas that do not receive projections from primary sensory areas project back to them nevertheless (Rockland and Van Hoesen 1994). Additionally, feedback axons may project back to more than one early association cortex (Rockland and Drash 1996). The most dramatic instances of nonreciprocal feedback projections are those that arise from some of the endstations of cortical feedforward axons. For example, the subicular part of the hippocampal formation as well as layer IV of the entorhinal cortex have widespread projections to parts of the limbic, temporal and frontal lobes from which they do not receive an afferent (Rosene and Van Hoesen 1977; Kosel et al. 1982). The same can be said for the amygdala and, in fact, projections directly to the primary auditory and visual cortices have been described (Amaral and Price 1984). Finally, the neurons that form the nucleus basalis of Meynert project to all of the cerebral cortex (Mesulam et al. 1983) despite receiving input from only the distal-most association cortices of the temporal and frontal lobes and the anterior insula (Mesulam and Mufson 1984).

In summary, it could be argued that feedforward projections are skewed in the direction of synthesis and send on to their endstations a highly abstract form of outcomes. Feedback connections could be viewed as magnifying the process of synthesis. In the early association cortices there appears to be a near one-to-one, or directly reciprocal, feedforward-feedback relationship. However, distal association areas project directly back to areas that only indirectly influence them and the endstations project back to all areas in the stream whether they received a direct projection from them or not. In a generic sense it is plausible to believe that feedback axons impart something akin to "meaning" or "novelty" to cortical processing and, without this element, new learning cannot occur.

Feedforward Endstations and Feedback Origins in Alzheimer's disease

Entorhinal/Hippocampal Cortex

It is arguable that cortical feedforward axons retrace in part the major events that occurred in the evolution of the cerebral cortex. That is, they course sequentially from the newest, or most highly evolved cortex, the primary sensory areas, to the oldest and least elaborate cortex found in the hippocampal formation. The intermediate cortices, or the association areas, are part of this stepwise sequence but have substantial interconnectivity among themselves. A simple rule could govern this sequence, such that each newly evolved cortical area projects to its immediate progenitor. Feedback connections follow the inverse of this rule, projecting back to their newly differentiated progeny. However, as discussed above, this would appear substantially looser, such that the older progenitor may project to and influence progeny several generations beyond its first offspring.

Some validity for the notions discussed above can be gleaned by examining the cortical connections of the entorhinal cortex. This cortex receives a host of cortical projections from areas that surround it and other parts of the limbic lobe, but its strongest input arises from the perirhinal and posterior parahippocampal cortex (Fig. 1), two areas of cortex that receive extensive projections from the distal visual association areas (Van Hoesen 1982; Insausti et al. 1987; Suzuki and Amaral 1994). Layers II and III of the entorhinal cortex give rise to the perforant pathway, which ends on the distal dendrites of dentate gyrus granule cells and hippocampal pyramidal neurons (Fig. 2). Intrinsic connectivity within the hippocampal formation leads to strong projections to its CA1 and subicular sec-

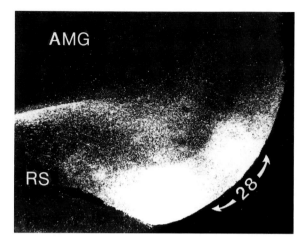

Fig. 1. A dark field photomicrograph of a coronal section through the monkey entorhinal cortex (area 28) with autoradiography demonstrating axonal and terminal labeling (white) following an injection of tritiated amino acids into the visual association cortex area TF (AMG, amygdala; RS, rhinal sulcus)

tors, and these give rise to long feedback axons back to the cortex including layer IV of the entorhinal cortex which has similar long cortical projections (Rosene and Van Hoesen 1977; Kosel et al. 1982).

It is of interest that in Alzheimer's disease, the perirhinal and entorhinal cortices are damaged heavily by neurofibrillary tangles (Fig. 3), compromising this endstation of feedforward cortical axons. Additionally, neurofibrillary tangles

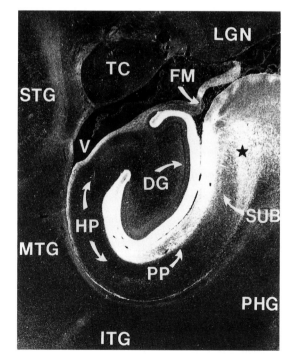

Fig. 2. A dark field photomicrograph of a coronal section through the monkey hippocampal formation with autoradiography demonstrating axonal and terminal labeling (white) over the perforant pathway (PP) terminal zone following an injection of tritiated amino acids into the entorhinal cortex. The star marks the position of the labeled angular bundle and axons perforating through the subiculum (SUB). Note, the dense terminal labeling over both the dentate gyrus (DG) and hippocampus (HP). FM, fimbria; ITG, inferior temporal gyrus; LGN, lateral geniculate nucleus; MTG, middle temporal gyrus; PHG, parahippocampal gyrus; STG, superior temporal gyrus; TC, tail of caudate nucleus; V, lateral ventricle)

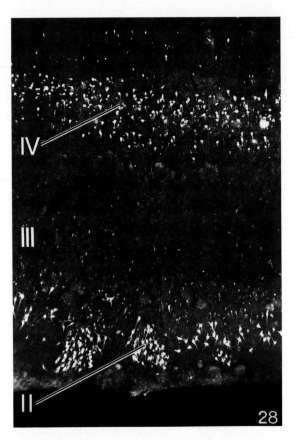

Fig. 3. A thioflavin S-stained coronal section through the entorhinal cortex (area 28) in Alzheimer's disease showing neurofibrillary tangles in layers II and IV.

invest subicular (Fig. 4) and entorhinal layer IV (Fig. 3) neurons, compromising the initial step of feedback projections to the cortex (Hyman et al. 1984, 1986, 1988; Braak and Braak, 1985, 1991, 1992; Arnold et al., 1991; Van Hoesen et al., 1991; Vermersch et al., 1992).

Amygdala

The cerebral cortex provides the greatest source of input to the amygdala. Corticoamygdaloid axons are highly organized such that different cytoarchitectural areas project preferentially to discrete amygdaloid nuclei and their subdivisions (Herzog and Van Hoesen 1976; Van Hoesen 1981; Amaral et al. 1992). For example, the distal visual association cortices project to the dorsal and lateral parts of the lateral amygdaloid nucleus and to the dorsal part of the laterobasal nucleus. In contrast, the distal auditory association cortex projects to the ventral and lateral parts of the lateral nuclei. The temporal polar cortex projects to the medial part of the lateral amygdaloid nucleus and to the magnocellular part of the

Fig. 4. Neurofibrillary tangles (NFTs) and a neuritic plaque (NP) in the subiculum (SUB) in Alzheimer's disease, stained with thioflavin S.

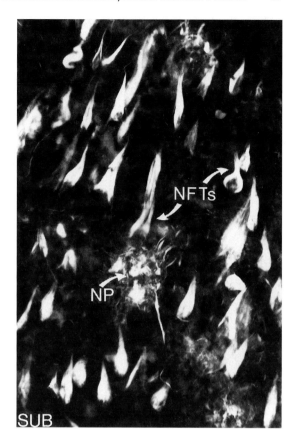

accessory basal nucleus (Fig. 5). Perirhinal and posterior parahippocampal areas project to the mediobasal nucleus and the parvocellular part of the accessory basal nucleus, as does the subicular cortex of the hippocampal formation.

All of the corticoamygdaloid projections are powerful corticofugal neural systems and must be viewed as another endstation of feedforward cortical axons in line multisynaptically with the output of the primary sensory areas (Fig. 6). Nearly all of the anterior temporal distal association areas that give origin to these projections are damaged heavily by neurofibrillary tangles and neuritic plaques in Alzheimer's disease (Arnold et al. 1991; Braak and Braak 1991). Additionally, Alzheimer's disease pathology is present heavily in many amygdaloid nuclei (Kromer-Vogt et al. 1990), which compromises strong amygdalocortical feedback axons that project back to both distal and early association cortex as well as to the primary sensory areas for the auditory and visual modalities (Amaral et al. 1992). In short, this is another instance where the pathology of Alzheimer's disease damages both the endstation of feedforward axons and the initial origin of cortical feedback axons.

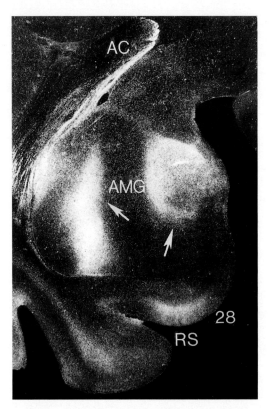

Fig. 5. A dark field photomicrograph of a coronal section through the monkey amygdala (AMG) with autoradiography demonstrating axonal and terminal labeling (white) over the lateral nucleus (left arrow) and accessory basal nucleus (right arrow) following a tritiated amino acid injection into the temporal polar cortex. AC, anterior commissure; RS, rhinal sulcus.

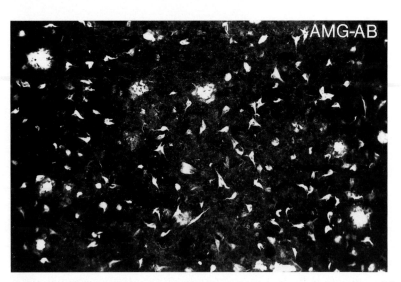

Fig. 6. Neurofibrillary tangles and neuritic plaques in the accessory basal nucleus of the amygdala (AMG-AB) in Alzheimer's disease, stained with thioflavin S.

Nucleus Basalis of Meynert/Substantia Innominata

The nucleus basalis of Meynert is a collection of large hyperchromatic neurons that occupy in part the substantia innominata of the basal forebrain (Kievit and Kuypers 1975; Divac 1975; Jones et al. 1976; Mesulam and Van Hoesen 1976) ventral to the anterior commissure and globus pallidus. Their cholinergic chemistry sets them apart (Mesulam et al. 1983) but also reveals that scattered groups of neurons extend like tentacles anteriorly into the diagonal bands of Broca and septal region, posteriorly toward the hypothalamus and midbrain and laterally to the globus pallidus and amygdala. The major output of the nucleus basalis of Meynert is to the cerebral cortex (Pearson et al. 1983; Mesulam et al. 1983; Wenk et al. 1980; Fibiger 1982) where it provides cholinergic innervation (Fig. 7). However, and importantly, this nucleus also projects to the thalamic reticular nucleus (Levey at al. 1987; Buzsaki et al. 1988; Asanuma 1989). This places the nucleus basalis of Meynert in a position to influence the cortex indirectly as well, because the thalamic reticular nucleus governs thalamic transmission via intrinsic thalamic inhibitory connections (Jones 1975).

Curiously, the nucleus basalis of Meynert receives cortical projections from only a small fraction of the cortex it projects to (Fig. 8). Notable sources of input are from the anterior insular, medial frontal, temporal polar, orbitofrontal and entorhinal cortex (Mesulam and Mufson 1984). A strong input is also received from the amygdala (Price and Amaral 1981), and the hippocampal formation projects strongly to septal and diagonal band cholinergic neurons (Fig. 9). These are all endstations for multisynaptic streams of cortical axons. It could be argued that the nucleus basalis of Meynert influences all levels of cortical processing, either directly or indirectly via the thalamic reticular nucleus, but is influenced

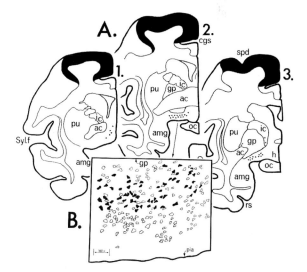

Fig. 7. Coronal sections 1–3 in A depict the location of a horseradish peroxidase injection (darkened area) in cortical areas 4 and 6. Retrogradely labeled neurons in the nucleus basalis of Meynert are shown by triangles in A and as darkened profiles in B. ac, anterior commissure; amg, amygdala; cgs, cingulate sulcus; gp, globus pallidus; h, hypothalamus; ic, internal capsula; oc, optic chiasm; pu, putamen; spd, superior precentral dimple; rs, rhinal sulcus; Syl. f, Sylvian fissure.)

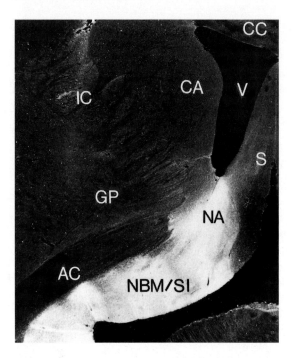

Fig. 8. A dark field photomicrograph of a coronal section through the monkey basal ganglia and basal forebrain with autoradiography demonstrating axonal and terminal labeling (white) over the nucleus basalis of Meynert of the substantia innominata (NBM/SI) and nucleus accumbens (NA) after an injection of tritiated amino acids into the amygdala. The labeled axon terminals are known to contact nucleus basalis of Meynert cholinergic neurons. AC, anterior commissure; CA, caudate nucleus; CC, corpus callosum; GP, globus pallidus; IC, internal capsule; S, septum; V, lateral ventricle.)

only by cortical endstations after the whole sequence of corticocortical processing streams has been traced.

In Alzheimer's disease, the cortical areas that project to the nucleus basalis of Meynert and the other cholinergic neurons of the basal forebrain are heavily damaged (Arnold et al. 1991; Braak and Braak 1991). Likewise, the cholinergic enzymes in the cortex are diminished (Davies and Maloney 1976) and the nucleus basalis of Meynert is damaged (Whitehouse et al. 1981, 1982; Arendt et al. 1983, 1985; Wilcock et al. 1983; Tourtellotte et al. 1989, Geula and Mesulam 1996). Thus, like the entorhinal/hippocampal cortex and the amygdala, the nucleus basalis of Meynert and its input/output relationships provide another example in Alzheimer's disease where the endstations of cortical feedforward axons and the origin of cortical feedback axons are damaged heavily.

Summary and Conclusions

There can be little argument that Alzheimer's disease is characterized neuroanatomically by profound changes in cortical connectivity and that long association systems are compromised greatly. The essential neurons for these systems either contain neurofibrillary tangles (Hof and Morrison 1990, 1994; Hof et al. 1990) or they are diminished in number (Coleman and Flood 1987; Hyman et al. 1995; Gómez-Isla et al. 1996, or their probable synapses are altered (Hamos et al. 1989;

Fig. 9. A dark field photomicrograph of a coronal section through the monkey basal ganglia and basal forebrain with autoradiography demonstrating axonal and terminal labeling (white) over the septum/diagonal band (S) nuclei and the nucleus accumbens (NA) following an injection of tritiated amino acids into the CA1/subiculum sectors of the hippocampal formation. Many of these axon terminals contact cholinergic neurons in the medial septal nucleus and vertical limb of the diagonal band of Broca. CA, caudate nucleus; CC, corpus callosum; GP, globus pallidus; IC, internal capsule; V, lateral ventricle.)

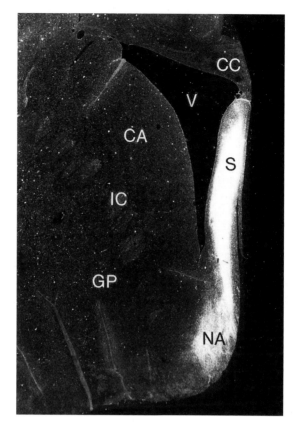

DeKosky and Scheff 1990; Masliah et al. 1991; Terry et al. 1991). Since long association cortical systems have long been linked to cognitive processes (Van Hoesen 1993), it is logical to believe that their destruction is coupled tightly to cognitive alterations and dementia. However, this viewpoint is not always comfortable. In any large collection of Alzheimer's disease brains, many can be found to have reasonable brain weight, negligible atrophy and limited neurofibrillary tangles and neuritic plaques, yet still have confirmed clinical dementia for several years. With regard to the association cortices, the same brains may differ greatly from a matched set where brain weight is low, atrophy is conspicuous and neurofibrillary tangles and neuritic plaques are abundant. It has been argued that the communality between such brains may lie in the invariant pathological changes in limbic structures that receive the endstation projections of cortical feedforward projections and give rise to widespread feedback cortical projections. These changes account for much of the memory impairment in Alzheimer's disease and may be related to the decline in executive control that tips the elderly patient into dementia. Pathological changes in the association cortices are more frequently than not part of the picture at endstage Alzheimer's disease, but it is arguable that they may be more related to the density of the cognitive change and not its trigger.

Finally, it is important to note that Damasio (1988, 1989, 1994) has challenged the notion of hierarchical processing in cortical neural systems for the good reason that distal association cortices can be destroyed in humans and only selectively alter cognition and knowledge. In his thinking feedback connections recreate or reactivate in the early association cortices the various elements that form a unified percept and bind them together. The limbic structures, where invariant pathology occurs in Alzheimer's disease, could alter this ability to reactivate these neuronal ensembles, even though the cortex in which they reside is relatively intact. Examination of Damasio's more modern model of cognition in relationship to Alzheimer's disease could greatly expand our understanding of the behavioral changes that accompany this disorder, and reduce the variance that accompanies dealing with it in a unified manner.

Acknowledgments

The author thanks Paul Reimann for aid with photomicroscopy and Patty Frantz for typing the manuscript. Supported by NS 14944 and NS PO1 19632.

References

Amaral DG, Price JL (1984) Amygdalo-cortical projections in the monkey *(Macaca fasciularis).* J Comp Neurol 230: 465–496

Amaral DG, Price JL, Pitkänen A, Carmichael ST (1992) Anatomical organization of the primate amygdaloid complex. In: Aggleton JP (ed.) The amygdala. Wiley-Liss, New York, pp 1–66

Arendt T, Bigl V, Arendt A, Tennstedt A (1983) Loss of neurons in the nucleus basalis of Meynert in Alzheimer's disease, paralysis agitans and Korsakoff's disease. Acta Neuropathol (Berl) 61: 101–108

Arendt T, Bigl V, Arendt A, Tennstedt A (1985) Neuronal loss in different parts of the nucleus basalis is related to neuritic plaque formation in cortical target areas in Alzheimer's disease. Neuroscience 14: 1–14

Arnold S, Hyman BT, Flory J, Damasio AR, Van Hoesen GW (1991) The topographical and neuoranatomical distribution of neurofibrillary tangles and neuritic plaques in the central cortex of patients with Alzheimer's disease. Cereb Cortex 1: 103–116

Asanuma C (1989) Axonal arborizations of a magnocellular basal nucleus input and their relation to the neurons in the thalamic reticular nucleus of rats. Proc Natl Acad Sci USA 86: 4746–4750

Barbas H (1986) Pattern in the laminar origin of corticocortical connections. J Comp Neurol 252: 415–422

Braak H, Braak E (1985) On areas of transition between entorhinal allocortex and temporal isocortex in the human brain. Normal morphology and lamina-specific pathology in Alzheimer's disease. Acta Neuropathol (Berl) 68: 325–332.

Braak H, Braak E (1991) Neuropathological stageing of Alzheimer-related changes. Acta Neuropathol 82: 239–259

Braak H, Braak E (1992) The human entorhinal cortex: normal morphology and lamina-specific pathology in various diseases. Neurosci Res 15: 6–31.

Buzsaki G, Bickford RG, Ponomareff G, Thal LJ, Mandel R, Gage FH (1988) Nucleus basalis and thalamic control of neocortical activity in the freely moving rat. J Neurosci 8: 4007–4026.

Coleman PD, Flood DG (1987) Neuron numbers and dendritic extent in normal aging and Alzheimer's disease. Neurobiol Aging 8: 521–545.

Damasio AR (1988) The brain binds entities and events by multiregional activation from convergence zones. Neural Comp 1: 123–132

Damasio AR (1989) The time-locked multiregional retroactivation: A systems level proposal for the neural substrates of recall and recognition. Cognition 33: 25–62.

Damasio AR (1994) Descartes' Error: Emotion, reason, and the human brain. Grosset/Putnam, New York

Davies P, Malone AJF (1976) Selective loss of central cholinergic neurons in Alzheimer's disease. Lancet 2: 1403

DeKosky ST, Scheff SW (1990) Synapse loss in frontal cortex biopsies in Alzheimer's disease: Correlation with cognitive severity. Ann Neurol 27: 457–464

Divac I (1975) Magnocellular nuclei of the basal forebrain project to neocortex, brain stem and olfactory bulb: Review of some functional correlates. Brain Res 93: 385–398

Felleman DJ, Van Essen DC (1991) Distributed hierarchical processing in the primate cerebral cortex. Cereb Cortex 1: 1–47

Fibiger HC (1982) The organization and some projections of cholinergic neurons of the mammalian forebrain. Brain Res Rev 4: 327–388

Galaburda AM, Pandya DN (1983) The intrinsic architectonic and connectional organization of the superior temporal region of the rhesus monkey. J Comp Neurol 221: 169–184

Geschwind N (1965) Disconnection syndromes in animals and man. Brain 88: 237–294

Geula C, Mesulam M-M (1996) Systematic regional variations in the loss of cortical cholinergic fibers in Alzheimer's disease. Cereb Cortex 6: 165–177

Goldman-Rakic PS (1988) Topography of cognition: Parallel distributed networks in primate association cortex. Ann Rev Neurosci 11: 137–156

Gómez-Isla T, Price JL, McKeel DW, Morris JC, Growdon JH, Hyman BT (1996) Profound loss of layer II entorhinal cortex neurons occurs in very mild Alzheimer's disease. J Neurosc, 16: 4491–4500

Gross CG (1992) Representation of visual stimuli in inferior temporal cortex. Phil Trans R Soc Lond 335: 3–10

Hamos JE, Degennaro LJ, Drachman DA (1989) Synaptic loss in Alzheimer's disease and other dementias. Neurology 39: 355–361

Herzog AG, Van Hoesen GW (1976) Temporal neocortical afferent connections to the amygdala in the rhesus monkey. Brain Res 115: 57–69

Hof PR, Morrison JH (1994) The cellular basis of cortical disconnection in Alzheimer disease and related dementing conditions. In: Terry RD, Katzman R, Bick KL (eds.) Alzheimer disease. Raven Press, New York, pp 197–229

Hof PR, Morrison JH (1990) Quantitative analysis of a vulnerable subset of pyramidal neurons in Alzheimer's disease: II. Primary and secondary visual cortex. J Comp Neurol 301: 55–64

Hof PR, Cox K, Morrison JH (1990) Quantitative analysis of a vulnerable subset of pyramidal neurons in Alzheimer's disease: I. Superior frontal and inferior temporal cortex. J Comp Neurol 301: 44–54

Hyman BT, Van Hoesen GW, Damasio AR, Barnes CL (1984) Alzheimer's disease cell specific pathology isolates the hippocampal formation. Science 225: 1168–1170

Hyman BT, Van Hoesen GW, Kromer LJ, Damasio AR (1986) Perforant pathway changes and the memory impairment of Alzheimer's disease. Ann Neurol 20: 472–481

Hyman BT, Kromer LJ, Van Hoesen GW (1988) A direct demonstration of the perforant pathway terminal zone in Alzheimer's disease using the monoclonal antibody Alz-50. Brain Res 450: 392–397

Hyman BT, West HL, Gómez-Isla T, Mui S (1995) Quantitative neuropathology in Alzheimer's disease: Neuronal loss in high-order association cortex parallels dementia. In: Iqbal K, Mortimer JA, Winblad B, Wisniewski HM (eds.) Research advances in Alzheimer's disease and related disorders. John Wiley & Sons, New York, pp 453–460

Insausti R, Amaral DG, Cowan WM (1987) The entorhinal cortex of the monkey. II. Cortical afferents. J Comp Neurol 264: 326–355

Jones EG (1975) Some aspects of the organization of the thalamic reticular complex. J Comp Neurol 162: 285–308

Jones EG (1990) Determinants of the cytoarchitecture of the cerebral cortex. In: Edelman GM, Gall WE, Cowan WM (eds.) Signal and sense. Wiley-Liss, New York, pp 3–49

Jones EG, Powell TPS (1970) An anatomical study of converging sensory pathways within the cerebral cortex of the monkey. Brain 93: 37–56

Jones EG, Burton H, Saper CB, Swanson LW (1976) Midbrain, diencephalic and cortical relationships of the basal nucleus of Meynert and associated structures in primates. J Comp Neurol 167: 385–420

Kievit J, Kuypers HGJM (1975) Basal forebrain and hypothalamic connections to the frontal and parietal cortex in the rhesus monkey. Science 187: 660–662

Kosel KC, Van Hoesen GW, Rosene DL (1982) Non-hippocampal cortical projections from the entorhinal cortex in the rat and rhesus monkey. Brain Res 244: 201–213

Kromer Vogt LJ, Hyman BT, Van Hoesen GW, Damasio AR (1990) Pathological alterations in the amygdala in Alzheimer's disease. Neuroscience 37: 377–385

Levey AI, Hallanger AE, Wainer BH (1987) Cholinergic nucleus basalis neurons may influence the cortex via the thalamus. Neurosci Lett 74: 7–13

Masliah E, Terry RD, Alford M, De Teresa R, Hansen LA (1991) Cortical and subcortical patterns of synaptophysin-like immunoreactivity in Alzheimer's disease. Am J Pathol 138: 235–246

Mesulam M-M, Mufson EJ (1984) Neural inputs into the nucleus basalis of the substantia innominata (Ch. 4) in the rhesus monkey. Brain Res 107: 253–274

Mesulam M-M, Van Hoesen GW (1976) Acetylcholinesterase containing basal forebrain neurons in the rhesus monkey project to neocortex. Brain Res 109: 152–157

Mesulam M-M, Mufson EJ, Levey AI, Wainer BH (1983) Cholinergic innervation of cortex by the basal forebrain: Cytochemistry and cortical connections of the septal area, diagonal band nuclei, nucleus basalis (subsantia innominata), and hypothalamus in the rhesus monkey. J Comp Neurol 214: 170–197

Pandya DN, Kuypers HGJM (1969) Cortico-cortical connections in the rhesus monkey. Brain Res. 13: 13–36

Pandya DN, Yeterian EH (1985) Architecture and connections of cortical association areas. Cereb Cortex 4: 3–61

Pearson RCA, Gather KC, Bridal P, Power TPS (1983) The projection of the basal nucleus of Meynert upon the neocortex in the monkey. Brain Res 259: 132–136

Price JL, Amaral DG (1981) An autoradiographic study of the projections of the central nucleus of the monkey amygdala. J Neurosci 1: 1242–1259

Rockland KS, Pandya DN (1979) Laminar origins and termination of cortical connections of the occipital lobe in the rhesus monkey. Brain Res 179: 3–20

Rockland KS, Pandya DN (1981) Cortical connections of the occipital lobe in the rhesus monkey: Interconnections between areas 17, 18, 19 and the superior temporal sulcus. Brain Res 212: 249–270

Rockland KS, Drash GW (1996) Collateralized divergent feedback connections that target multiple cortical areas. J Comp Neurol 373: 529–548

Rockland KS, Van Hoesen GW (1994) Direct temporal-occipital feedback connections to striate cortex (V1) in the Macaque monkey. Cereb Cortex 4: 300–313

Rosene DL, Van Hoesen GW (1977) Hippocampal efferents reach widespread areas of cerebral cortex and amygdala in the rhesus monkey. Science 198: 315–317

Suzuki WA, Amaral DG (1994) Topographic organization of the reciprocal connections between the monkey entorhinal cortex and the perirhinal and parahippocampal cortices. J Neurosci 14: 1856–1877

Terry RD, Masliah E, Salmon DP (1991) Physical basis of cognitive alterations in Alzheimer's disease: Synapse loss is the major correlate of cognitive impairment. Ann Neurol 41: 572–580

Tourtellotte WG, Van Hoesen GW, Hyman BT, Tikoo RK, Damasio AR (1989) Alz-50 immunoreactivity in the thalamic reticular nucleus in Alzheimer's disease. Brain Res 515: 227–234

Van Essen DC, Felleman DJ, DeYoe EA, Olavarria J, Knierim J (1990) Modular and hierarchical organization of extrastriate visual cortex in the Macaque monkey. Cold Spring Harbor Symp Quant Biol 3: 679–696

Van Hoesen GW (1981) The differential distribution, diversity and sprouting of cortical projections to the amygdala in the rhesus monkey. In: Ben-Ari Y (ed) The amygdaloid complex. Amsterdam, Elsevier/North Holland, pp 77–90

Van Hoesen GW (1982) The parahippocampal gyrus. Trends Neurosci 5: 345–350

Van Hoesen GW (1993) The modern concept of association cortex. Curr Opinion Neurobiol 3: 150–154

Van Hoesen GW, Hyman BT, Damasio AR (1991) Entorhinal cortex pathology in Alzheimer's disease. Hippocampus 1: 1–8

Vermersch P, Frigard B, Delacourte A (1992) Mapping of neurofibrillary degeneration in Alzheimer's disease – evaluation of heterogeneity using the quantification of abnormal Tau proteins. Acta Neuropathologica, 85: 48–54

Wenk H, Bigl V, Meyer V (1980) Cholinergic projections from magnocellular nuclei of the basal forebrain to cortical areas in rats. Brain Res Rev 2: 295–316

Whitehouse PJ, Price DL, Clark AW, Coyle JT, De Long MR (1981) Alzheimer's disease: Evidence for selective loss of cholinergic neurons in the nucleus basalis. Ann Neurol 10: 122–126

Whitehouse PJ, Price DL, Struble RG, Clark AW, Coyle JT, DeLong MR (1982) Alzheimer's disease and senile dementia: loss of neurons in the basal forebrain. Science 215: 1237–1239

Wilcock GK, Esiri MM, Bowen DM, Smith CCT (1983) The nucleus basalis in Alzheimer's disease: cell counts and cortical biochemistry. Neuropathol Appl Neurobiol 9: 175–179

Zeki SM (1975) The functional organization of projections from striate to prestriate visual cortex in the rhesus monkey. Cold Spring Harbor Symp Quant Biol 40: 591–600

Zeki S (1990) Functional specialization in the visual cortex: The generation of separate constructs and their multistage integration. In: Edelman GM, Gall WE, Cowan WM (eds.). Signal and sense. Wiley-Liss, New York, pp 85–130

Zeki S (1993) The visual association cortex. Curr Opinion Neurobiol 3: 155–159

Zeki S, Shipp S (1988) The functional logic of cortical connections. Nature 335: 311–317

Plaques and Tangles: Where and When?

C. Duyckaerts[*]*, M.-A. Colle, M. Bennecib, Y. Grignon, T. Uchihara, J.-J. Hauw*

Introduction

Alzheimer pathology includes two main aspects (Duyckaerts et al. 1995): the extracellular deposition of β amyloid (Aβ) and the intracellular accumulation of abnormally phosphorylated tau protein. The relationship between these two markers and their link with dementia are still a matter of debate (Selkoe 1991; van de Nes et al. 1994). Tau pathology appears to be more closely associated with dementia than Aβ deposition (Berg et al. 1993; McKee et al. 1991; Wilcock and Esiri 1982) but Aβ could play the role of a trigger (Hardy 1992) and be the cause of a "cascade," leading finally to the formation of neurofibrillary tangles and neuropil threads.

Cases and Methods

To elucidate the relationship between the pathological markers (Aβ and tau) and the progression of dementia, we relied on a cohort of patients who had been living in a long-stay hospital. They had all been admitted in the same geriatric ward from June 1983 to April 1984. One hundred and thirty-six women were examined. Seventy-seven were excluded from the study because possible causes of dementia other than Alzheimer's disease were found (strokes, Parkinson's disease, psychiatric diseases, alcoholism) or because severe sensory deficits (deafness, blindness) precluded adequate testing. An assessment of intellectual status by the Blessed Test Score (BTS) was made in 59 patients. Among those, 31 patients died. Two cases were excluded because of ischemic lesions. Complete microscopical data were available on 28 cases. The sampling was planned to cover the major types of cortex: primary sensory (visual primary cortex, area 17), associative unimodal (area 22 in the temporal lobe), associative multimodal (frontal, area 9; parietal, area 40) and limbic (area TF-TH according to van Hoesen (1982) in the parahippocampal gyrus and hippocampus with subiculum).

---

[*] Laboratoire de Neuropathologie R. Escourolle, Hôpital de La Salpêtrière, 47 Blvd de l'Hôpital, 75651 Paris Cedex 13

B. T. Hyman / C. Duyckaerts / Y. Christen (Eds.)
Connections, Cognition, and Alzheimer's Disease
© Springer-Verlag Berlin Heidelberg 1997

The staining included immunohistochemistry with a polyclonal rabbit antibody against tau protein (diluted 1/1000) from Dako® (Glostrup, Denmark) and with a monoclonal antibody against residues 8–17 of the Aβ peptide, also from Dako® (Glostrup, Denmark).

Tau positive neurofibrillary tangles and senile plaques, as well as Aβ deposits, were counted in columns of contiguous microscopic fields, covering the entire thickness of the cortex from the pial surface down to the white matter (Duyckaerts et al. 1986). Results were expressed in n/mm^2. Complete maps of neurofibrillary tangles in the hippocampal sample and of tau-positive neurofibrillary tangles and senile plaques in the supramarginal gyrus were obtained with a semiautomatic system, provided by Biocom® (Les Ulis, France). The maps were analyzed with a computer program developed in this laboratory (Duyckaerts et al. 1994), based on a division (i.e., a "tessellation") of the map into small areas including only one point (plaque, tangle or neuronal profile, according to the map). Those polygonal areas were inversely proportional to the numerical density, i.e., to the number of points per unit area: when this density was high, the areas were small and *vice versa*.

Results

The densities of the tangles in area TF-TH and in the subiculum were highly correlated with the intellectual status, as estimated by the BTS (r = 0.68; p < 0.0001). We tried to determine which subareas were initially involved by focusing on the maps of tangles that we obtained in the least affected cases. In those maps, we found a focus of lesions in the CA1-subiculum sector, as expected, but also a second focus in the region CA2–CA3 (unpublished data; P. Fouillard, in preparation; Fig. 1). Lesions appeared to be grouped in small clusters that coalesced into larger foci.

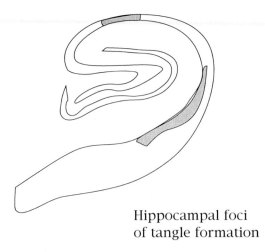

Hippocampal foci
of tangle formation

Fig. 1. Diagram showing the two foci of tangle formation that were detected in the hippocampus of the least affected cases of the series. The most prominent one involves the subicular/CA1 region, the second focus is located in the CA2 region

The correlations between lesion density and dementia were also significant for the tangles in the isocortical areas, but the regression curves showed that these relationships were not linear. These curves revealed that the presence of tangles was determined by clinical "thresholds:" tangles appeared to be present in a particular area only when the intellectual status of the patient had reached a given severity. The clinical thresholds were different for the various isocortical areas that we studied, with the consequence that, at a given intellectual status only a given number of areas were involved (Duyckaerts et al., 1997). To test this hypothesis, we counted, for each case, the number of areas that had at least one tangle. This number ("diffusion index") theoretically varied from 0 (no area involved) to 6 (all the areas involved in a sample including subiculum, area TF-TH, areas 40, 22, 9, 17). Correlation between this index and the intellectual status was high ($r = -0.93$; $p < 0.0001$). The order of involvement of the cortical areas was not random but followed a hierarchy that we tried to elucidate by computing a "hierarchical index:" we hypothesized that the involvement of a given area was *always* associated with the involvement of areas that occupied a lower order in the hierarchy. The area having the highest hierarchical order was thus the least often affected one: among the six areas that we studied, it happened to be the visual cortex (area 17). We isolated the cases with involvement of this cortex and confirmed that all of the other areas also contained tangles (if sometimes only a few). We then studied the cases with no tangles in area 17 and looked for the least affected cortex. Frontal area 9 was, in this way, taken as the next area in the hierarchy. By repeating the procedure, we found the following order of involvement: subiculum, areas TF-TH, 22, 40, 9, 17. A hierarchical index of involvement was attributed to each case by successively examining the six aforementioned areas and taking into account only the cortex of the highest order with at least one tangle. Cases with tangles only in the subiculum received index 1; those with tangles in areas TF-TH received index 2, etc... (see Table 1).

Table 1. Hierarchical index. The hierarchical index (Hier. index) depends on the topography of the tangles and is independent from their number. X means at least 1 neurofibrillary tangle (NFT); 0 means no tangles. A case with neurofibrillary tangles only in the subiculum is, for example, graded 1; a case with tangles in area 9 and no tangles in area 17 is graded 5[a]

Area	Presence of at least 1 NFT					
Subiculum	X	0	0	0	0	0
TF-TH		X	0	0	0	0
22			X	0	0	0
40				X	0	0
9					X	0
17						X
Hier. Index	1	2	3	4	5	6

[a] TF-TH, areas of the parahippocampal gyrus according to von Bonin (van Hoesen 1982); 22, Brodmann area 22 in the first temporal gyrus (unimodal associative); 40, Brodmann area 40, supramarginal gyrus (multimodal associative); 9, Brodmann area 9 in the midfrontal gyrus (multimodal associative); 17, Brodmann area 17, visual primary cortex

Table 2. Observed distribution of the hierarchical index in our series of 28 cases. The hierarchical index (area of the highest order exhibiting at least one tangle) is marked by X. When an area of hierarchical order n contained neurofibrillary tangles (NFT), areas of lower order were also involved in all the cases of our series

Area	Presence of at least 1 NFT					
Subiculum	X	0	0	0	0	0
TF-TH	x	X	0	0	0	0
22	x	x	X	0	0	0
40	x	x	x	X	0	0
9	x	x	x	x	X	0
17	x	x	x	x	x	X
Hier. Index	1	2	3	4	5	6
No. of cases.[a]	2	7	2	5	5	7

[a] No. of cases, number of cases at a given hierarchical index.

The hierarchical index was almost perfectly correlated with the diffusion index ($r = 0.994$; $p < 0.0001$). The presence of neurofibrillary tangles followed the order indicated in Table 2: when an area of hierarchical order n contained neurofibrillary tangles, all the areas of hierarchical order lower than n *also* contained tangles.

We followed the same procedure with Aβ deposits, either diffuse or focal, and only considered their presence, not their density. The number of areas with Aβ deposits was much higher than areas with tangles. Only two cases were found to be completely free of deposits. In 20 cases, six of six areas contained Aβ deposits, and in four cases, five of six areas were involved. The spared areas were usually limbic (hippocampus or parahippocampal gyrus did not contain Aβ deposits in seven cases). A fairly large number of samples contained Aβ deposits without neuritic pathologies. We could isolate 46 of those samples with "amyloid only" pathology. They contained diffuse, focal and perivascular deposits, some of them Congo red-positive; the senile plaques were devoid of neurites and contained only the amyloid core.

Discussion

These data suggest that counts of tangles in the limbic/paralimbic areas provide information similar to the simple detection of tangles in a "hierarchy" of cortices. The term hierarchy is used here to mean that the presence of tangles has a different meaning depending on the involved cortices: although common in the subiculum, they are present in the visual area only in the most affected cases. The cortical areas may thus be ranked in such an order that the involvement of one area is always associated with the involvement of the areas of a lower order. This confirms qualitative data, leading to the pathological staging proposed by Braak and Braak (Braak and Braak 1991; Braak et al. 1993). It suggests that the isocortical stages themselves could probably be further subdivided according to the topography of the tangles in the various types of isocortical areas.

Aβ deposits and tangle (tau) pathology have a different impact on the limbic-paralimbic and isocortical areas. Neurofibrillary tangles are first seen in the former, Aβ pathologies in the latter. If the cascade hypothesis were true (at least in its simplest formulation), then one should always see Aβ deposition when tangles are present, even in brains where the neuritic pathology is confined to the limbic cortices. This does not seem to be true; some cases (two in this cohort) were apparently "amyloid" free, an observation already made by others (Bancher and Jellinger 1994), but a more extensive sampling might be necessary to ascertain definitely the absence of Aβ deposits in large areas of the isocortex.

The relationship between dementia and lesions has often been thought of as linear: the more abundant the lesions, the more severe the intellectual deficit (Blessed et al. 1968). In our cases, this model only applied to the neurofibrillary tangles in the limbic and paralimbic cortices (see diagram of Fig. 2), where the density of lesions had no threshold and no ceiling, or a very high one, barely visible in the most affected cases of the cohort. In the isocortex, the diffusion of neurofibrillary tangles was more related to the dementia than was their density in a particular area. Moreover, the diffusion of tangle pathology followed a "hierarchy-like" progression (see Fig. 2), with the presence of tangles in a given isocortical area being always associated with their occurrence in areas of lower order in the hierarchy. The disease seemed to progress by gradually "filling" low order areas with tangles before reaching those of a higher order. Aβ deposition did not follow as precisely the anatomy of the cortex as tangle pathology. The

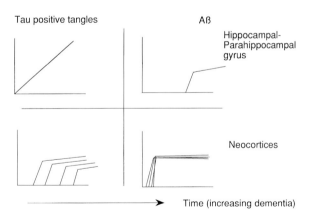

Fig. 2. Diagram illustrating the progression of the various markers according to the progression of dementia, as may be hypothesized from our data. Upper half, limbic/paralimbic cortex; lower half, isocortex; left half, tau positive tangles; right half, Aβ deposits. The density of tangles linearly increases with dementia in the hippocampal and parahippocampal gyri and gradually in the isocortex, where the presence of tangles seems to be determined by thresholds. Aβ deposits relatively spare the hippocampal and parahippocampal cortices and could diffuse to all of the isocortical areas within a short period of time (see text)

number of cases where Aβ deposits (see Fig. 2) were present in only a few areas was small, suggesting that numerous areas were involved at once. Moreover, deposition occurred in the walls of subarachnoid vessels and in the white matter (Uchihara et al. 1995 b) at a distance from neurons. The presence of Congo red-positive deposits, although more common in tangle-rich areas, was also noticed in tangle-free cortices. These observations are further evidence that amyloid transformation of Aβ may occur without degenerating neurites and is probably independent of neurons. Macrophages, as already stated (Hauw et al. 1988; Uchihara et al. 1995 a), may be the cells that are responsible for the transformation of Aβ peptide into congophilic, β-pleated material. However, the classical senile plaques associate Aβ deposition and tau abnormal neurites within a single structure. The tau positive abnormal neurites were present in tangle-rich areas, pointing out that the link between classical senile plaques and neurofibrillary tangles is probably strong (Probst et al. 1989). Areas devoid of tangles, such as the cerebellum and the striatum, are also devoid of neuritic plaques. These data may be taken as evidence that the abnormal neurites of the plaque crown come from tangle-bearing neurons.

What is the initial event: the entorhinal tangle or the isocortical deposit of Aβ peptide? The APP mutations leading to Alzheimer's disease (Goate et al. 1991) favor the latter hypothesis. However, the morphological evidence is still lacking: for the neuropathologist the tangle remains the most reliable and early marker of Alzheimer's disease.

Acknowledgments

This work is dedicated to the patients who gave a part of themselves for a better understanding of the brain and of its diseases.

We gratefully acknowledge the collaboration of the clinicians who examined the patients over the years, and particularly, of professor François Piette, Hôpital Charles Foix, Ivry.

This study would not have been possible without the expert help of the technical staff of Raymond Escourolle Laboratory, La Salpêtrière Hospital, Paris.

References

Bancher C, Jellinger KA (1994) Neurofibrillary predominant form of Alzheimer's disease: a rare subtype in very old subjects. Acta Neuropathol (Berlin) 88: 565–570

Berg L, McKeel DW, Miller P, Baty J, Morris JC (1993) Neuropathological indexes of Alzheimer's disease in demented and nondemented persons aged 80 years and older. Arch Neurol 50: 349–358

Blessed G, Tomlinson BE, Roth M (1968) The association between quantitative measures of dementia and of senile change in the cerebral grey matter of elderly subjects. Brit J Psychiat 114: 797–811

Braak H, Braak E (1991) Neuropathological stageing of Alzheimer-related changes. Acta Neuropathol (Berlin) 82: 239–259

Braak H, Duyckaerts C, Braak E, Piette F (1993) Neuropathological staging of Alzheimer-related changes correlates with psychometrically assessed intellectual status. In: Corain B, Iqbal K, Nicolini M, Winblad B, Wisniewski H, Zatta P (eds) Alzheimer's disease: advances in clinical and basic research. Chichester, John Wiley & Sons, pp 131–137

Duyckaerts C, Hauw J-J, Bastenaire F, Piette F, Poulain C, Rainsard V, Javoy-Agid F, Berthaix P (1986) Laminar distribution of neocortical plaques in senile dementia of the Alzheimer type. Acta Neuropath (Berlin) 70: 249–256

Duyckaerts C, Godefroy G, Hauw J-J (1994) Evaluation of neuronal numerical density by Dirichlet tessellation. J Neurosci Meth 51: 47–69

Duyckaerts C, Delaère P, He Y, Camilleri S, Braak H, Piette F, Hauw J-J (1995) The relative merits of tau- and amyloid markers in the neuropathology of Alzheimer's disease. In: Bergener M, Finkel SI (eds) Treating Alzheimer's and other dementias. New York: Springer, pp 81–89

Duyckaerts C, Bennecib M, Grignon Y, Uchihara T, He Y, Piette F, Hauw J-J (1997) Modeling the relation between neurofibrillary tangles and intellectual status. Neurobiol Aging (in press)

Goate A, Chartier-Harlin MC, Mullan M, Brown J, Crawford F, Fidani L, Giuffra L, Haynes A, Irving N, James L, Mant R, Newton P, Rooke K, Roques P, Talbot C, Pericak-Vance M, Roses A, Williamson R, Rossor M, Owen M, Hardy J (1991) Segregation of a missense mutation in the amyloid precursor protein gene with familial Alzheimer's disease. Nature 349: 704–706

Hardy J (1992) An'anatomical cascade hypothesis' for Alzheimer's disease. Trends Neurosci 15: 200–201

Hauw J-J, Duyckaerts C, Delaère P, Chaunu MP (1988) Maladie d'Alzheimer, amyloïde, microglie et astrocytes. Rev Neurol (Paris) 144: 155–157

McKee AC, Kosik KS, Kowall NW (1991) Neuritic pathology and dementia in Alzheimer's disease. Ann Neurol 30: 156–165

Probst A, Anderton BH, Brion JP, Ulrich J (1989) Senile plaque neurites fail to demonstrate anti-paired helical filament and anti-microtubule-associated protein tau immunoreactive proteins in the absence of neurofibrillary tangles in the neocortex. Acta Neuropathol 77: 430–436

Selkoe DJ (1991) Amyloid protein and Alzheimer disease. Sci Am November, 68–78

Uchihara T, Duyckaerts C, He Y, Kobayashi K, Seilhan D, Amouyel P, Hauw J-J (1995a) ApoE immunoreactivity and microglial cells in Alzheimer's disease brain. Neurosci Lett 195: 5–8

Uchihara T, Kondo H, Akiyama H, Ikeda K (1995b) White matter amyloid in Alzheimer's disease brain. Acta Neuropathol 90: 51–56

van de Nes JAP, Kamphorst W, Swaab DF (1994) Arguments for and against the primary amyloid local induction hypothesis of the pathogenesis of Alzheimer's disease. Ann Psychiat 4: 95–111

van Hoesen G (1982) The parahippocampal gyrus. New observations regarding its cortical connections in the monkey. Trends Neurosci 5: 345–350

Wilcock GK, Esiri MM (1982) Plaques, tangles and dementia. A quantitative study. J Neurol Sci 57: 407–417

Cortical Mapping of Pathological Tau Proteins in Several Neurodegenerative Disorders

P. Vermersch[*], V. Buée-Scherrer, L. Buée, J. P. David, A. Wattez, N. Sergeant, P. R. Hof, Y. Agid, D. P. Perl, C. W. Olanow, Y. Robitaille, D. Gauvreau, H. Petit, A. Delacourte

Introduction

One of the main features of numerous neurodegenerative disorders is the presence of specific inclusions in the cerebral cortex. In most of these disorders, these inclusions correspond to the aggregation of abnormal filaments. Neurofibrillary tangles (NFT) are commonly found in Alzheimer's disease (AD), amyotrophic lateral sclerosis/parkinsonism-dementia complex of Guam (ALS/PDC), head injury, Hallervorden-Spatz disease and progressive supranuclear palsy (PSP). Pick's disease (PiD), a form of fronto-temporal dementia, is characterized by the presence of chromatolytic neurons and Pick bodies (PB; Brion et al. 1991). Despite many microscopic and ultrastructural differences, these inclusions share similar antigenic properties (Dickson et al. 1985; Delacourte et al. 1990; Hof et al. 1994a, b). Tau immunoreactivity is observed in NFT of most of these neurodegenerative disorders and anti-Tau antibodies immmunolabel PB (Hof et al. 1994a).

Using a Western blot method, we demonstrated that aggregated, phosphatase-resistant Tau proteins are specifically immunodetected in SDS brain extracts from patients with AD, PSP or PiD (Delacourte et al. 1990, 1996; Vermersch et al. 1994). These pathological Tau proteins (PTP), which are the basic components of neurofibrillary lesions and PB, are different from normal Tau proteins, which are soluble and rapidly dephosphorylated after death (Matsuo et al. 1995). PTP have a specific profile that is different in AD, PiD and PSP (Vermersch et al. 1994; Buée-Scherrer et al. 1996a; Delacourte et al. 1996). We demonstrated earlier that the amounts of PTP were closely related to the quantities of neurofibrillary lesions (Flament et al. 1990). These markers allowed us to map easily the distribution of the neurodegenerating process.

---

[*] Unité INSERM 422, Place de Verdun, 59045 Lille cedex, France

B. T. Hyman / C. Duyckaerts / Y. Christen (Eds.)
Connections, Cognition, and Alzheimer's Disease
© Springer-Verlag Berlin Heidelberg 1997

Pathological Tau Proteins in Cortical Dementias and in Normal Aging

Distribution of Pathological Tau Proteins in Alzheimer's Disease

At the electron microscopic level, the neurofibrillary lesions observed in the brain from patients with AD are constituted of very characteristic filaments called paired helical filaments (PHF). These PHF are present in cell bodies of degenerating neurons, in neuritic extensions (neuropil threads) and in dystrophic neurites of neuritic plaques (Braak and Braak 1988). In 1985, Brion et al. showed that NFT were labelled by antibodies against the microtubule associated protein Tau. Following this pioneer work, many teams corroborated that PHF are formed almost entirely of Tau proteins. The Tau proteins of PHF share many epitopes with normal Tau but also have several distinct properties, such as lower mobility on sodium dodecyl sulfate-polyacrylamide gel electrophoresis and a different isoelectric charge (Sergeant et al. 1995). Several groups reported that excessive phosphorylation was the major modification in these proteins (Grundke-Iqbal et al. 1986; Flament et al. 1989). In 1990, three highly phosphorylated Tau proteins were characterized and named Tau 55, 64 and 69, according to their molecular weights (Delacourte et al. 1990) also referred as A68 or PHF-Tau (Lee et al. 1991; Goedert et al. 1993; Fig. 1). This triplet is known to

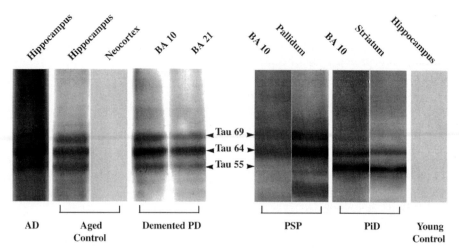

Fig. 1. Immunodetection of pathological Tau proteins with the monoclonal antibody AD2 on brain homogenates from patients with different neurodegenerative disorders and controls subjects. The pathological Tau triplet 55, 64 and 69 was detected in large amounts in the hippocampus from an AD patient. This triplet was also detected in the cortical areas (BA 10 and 21) from a demented PD patient and in the hippocampus from an aged non-demented control. A Tau doublet is observed in progressive supranuclear palsy (PSP; Tau 64 and 69) whereas another Tau doublet is observed in Pick's disease (PiD; Tau 55 and 64). There was no detection of PTP in homogenates from young controls or in the neocortex of aged non-demented controls.

be a specific and a reliable biochemical marker of the neurofibrillary degeneration of the Alzheimer type.

Using a Western blot method, we have been able to set up a reliable experimental protocol based upon an immunoblot analysis of PTP. We used the monoclonal antibody named AD2, which recognizes a phosphorylation-dependent Tau epitope (Buée-Scherrer et al. 1996b). The mapping was performed in almost all the Brodmann areas (BA) and the amounts of PTP were measured by densitometry, as described earlier (Vermersch et al. 1992a).

We noted that the amounts of PTP were uniform within each BA. Indeed the intensities of immunodetection were virtually identical in different parts of the same BA, whereas they were different from one Brodmann area to another (Vermersch et al. 1992a). This homogeneity within the same area argues for a relationship between the degenerating process and the cytoarchitectural regional scheme. In all cases, we found PTP in virtually all the cortical areas. As demonstrated using immunohistological tools (Lewis et al. 1987; Hof and Morrison 1990), BA 4 (primary motor cortex) was less involved and BA 17 (primary visual cortex) was sometimes spared (Vermersch et al. 1992a). The quantities of PTP were especially great in temporal neocortical and limbic areas and were higher in associative cortex than in primary sensory cortex. Amounts of PTP from occipital and frontal lobes differed strongly between patients as compared to the uniform degree of detection in the limbic, temporal and parietal lobes. Because all the brains were studied at the end stages of the disease, we cannot assume that the heterogeneity of the cortical mapping was related to specific clinical patterns.

Pathological Tau Proteins in Normal Aging

The presence of both senile plaques (SP) and NFT within the brain of non-demented old people has long been documented (Tomlinson et al. 1968). During aging, it is known that NFT appear first in the entorhinal cortex and then in hippocampus, whereas SP develop early in the frontal, temporal and neocortex (Braak and Braak 1991; Bouras et al. 1993). Using the same Western blot method, we investigated numerous brains from aged non-demented subjects and the abnormal Tau triplet was quantified in several cortical areas, including the entorhinal cortex (EC), hippocampus and BA 38, 20, 22, 35, 9, 44 and 39. PTP were detected in the EC of most of the controls aged over 70 years and in the hippocampus from controls aged over 80 years (Vermersch et al. 1995a; Fig. 1). However, the intensity of PTP immunodetection differs quantitatively compared to AD patients In some of the controls, PTP were also detected either in the isocortical BA 38 alone or also in BA 20 and BA 35. Soma cases, and especially those with Tau pathology in BA 38, 22 or 20, contained numerous SP in the neocortex. Few control subjects with Tau pathology restricted to the entorhinal cortex or hippocampus were almost devoid of SP in the neocortex. We failed to demonstrate a close relationship between the Tau pathology and the distribution of SP.

Due to the specificity of the Tau profile, we demonstrate that the degenerating process is typically of the Alzheimer type. The detection of lesions in the neocortex, even confined to a few temporal areas, raises the problem of "normal" and "pathological aging" (Giannakopoulos et al. 1995). Although these patients did not demonstrate any cognitive or memory impairment, it is possible that they represent what would constitute a group with increased risk for the development of AD. From a neuropathological point of view, these patients may represent an intermediate group between "normal aging", with lesions confined to the hippocampal formation, and "Alzheimer's disease", where lesions are more numerous and diffuse throughout the neocortex (Morris et al. 1991; Vermersch et al. 1992a, b; Arriagada et al. 1992). The Braak and Braak' study (1991), which demonstrated a specific pattern of lesion formation, supports this point of view. Some authors suggest that non-demented cases with relatively high NFT densities in the inferior temporal cortex without implication of other neocortical areas could represent a preclinical stage of AD (Hof et al. 1992a; Bouras et al. 1993). When restricted to the hippocampal area, the degenerating process represents a normal condition during aging and may be due to a selective vulnerability of a subset of neurons. Using a biochemical approach, our data also show that a slight involvement of other temporal neocortical areas is not systematically associated with the clinical signs of dementia.

Pathological Tau Proteins in Frontotemporal Dementias

Frontal lobe degeneration (FLD) is a common neurological disorder that has been recently characterized, despite the fact that it is the most common dementing disorder of the presenium after AD (Gustafson 1987). It belongs to the group of fronto-temporal dementia, with PiD. Both diseases have a similar "frontal" pathology, but PiD is easier to diagnose because of the characteristic Pick bodies found in the neurons from the fronto-temporal cortex and the hippocampal regions (Hof et al. 1994a). Occasionally the pathology of PiD is more extended, with lesions observed in parietal lobes. The absence of histological markers has hampered the identification of FLD since the morphological changes are not specific. They comprise gliosis, neuronal loss and spongiform changes mainly in the superficial cortical laminae.

Despite the fact that there are no specific histologic markers such as PB or neurofibrillary lesions, and therefore an absence of NFT and dystrophic neurites with PHF, we observed a slight amount of PTP in prefrontal biopsies from three FLD cases and in frontal BA 9, 10 and 32 and in temporal area 38 from an autopsic young FLD case (Vermersch et al. 1995b). The profile is a triplet as in AD, but smears of Tau aggregates, always observed by immunoblotting in AD, were not present. These proteins are likely in small aggregates of PTP that render them inaccessible to phosphatase activity generated during postmortem delays. It demonstrates that PTP can be detected without specific lesions observed at the optical level.

Recently, Tau proteins were analyzed by one- and two-dimensional gel electrophoresis, using a quantitative Western blot approach with the monoclonal antibody AD2 (Delacourte et al. 1996) in several PiD cases. Two major bands were detected, with a molecular weight of 55 and 64 kDa, in numerous areas of the frontal and temporal lobes (Fig. 1). The band corresponding to Tau 69 was present in very low amounts. In two of the five cases studied this Tau doublet was also observed in the parietal lobe, form example in the BA 39. The occipital lobe was always spared. Several subcortical nuclei were also involved such as the striatum, the amygdala, the nucleus basalis of Meynert, the pallidum, the thalamus, the substantia nigra and the locus coeruleus. The correlation between the presence of PB detected immunohistochemically and the PiD Tau doublet quantified biochemically was highly significant statistically (Delacourte et al. 1996). In some cases, the amounts of Tau 55 and 64 were higher in subcortical nuclei, for example in the striatum, than in the cortex. This finding demonstrates that PiD is both a cortical and a subcortical disorder.

Cortical Tau Pathology in Subcortical Dementias

Intensive investigations have demonstrated two principal patterns of neuropsychological deterioration within the dementias: 1) a cortical pattern with intellectual decline manifested typically by aphasia, amnesia, agnosia, acalculia and apraxia; and 2) a subcortical pattern characterized by slowing of cognition, memory disturbances, difficulty with complex intellectual tasks, such as strategy generation and problem solving, and disturbances of mood and affect (Cummings and Benson 1984). Anatomically, the cortical pattern is produced by diseases involving primarily, but not exclusively, the association cortex and the medial temporal lobes, whereas the subcortical pattern occurs in disorders with predominant involvement of basal ganglia, thalamus and brainstem structures. AD is the model for the cortical dementias. The dementia syndromes associated with Huntington's disease, PSP and Parkinson's disease (PD) exemplify the subcortical pattern of intellectual impairment. The relationship of dementia to cortical dysfunction is widely accepted, whereas the correlation between subcortical pathology involvement and the dementia remains controversial. Recent studies have shown significant pathological changes in the cortex from demented patients with such "subcortical patterns." Using our biochemical approach, we have investigated brains from patients with PSP or PD with various degrees of dementia.

Cortical Involvement in the Brain from Demented Patients with Parkinson's Disease

The prevalence of dementia in PD is greater than in the general population but controversy remains about the nature and neuropathological basis of cognitive deficits associated with the disease. The consequences of the dopamine decrease on the fronto-subcortical circuits, the extension of the Lewy body pathology to the cortex and the involvement of the mesolimbic pathway may explain this higher risk of dementia. Hakim and Mathieson (1979) found an increased number of AD lesions in the cerebral cortex of demented PD patients and suggested that the dementia was due to the association with AD. However, epidemiological studies have demonstrated that the two diseases co-occur no more frequently than predicted by chance in an elderly population.

To elucidate the neurochemical basis of the dementia of PD, we compared samples of cerebral cortex from 24 non-demented parkinsonian patients with various degree of dementia with samples from patients with AD and controls, using a quantitative Western blot analysis with the monoclonal antibody AD2 (Vermersch et al. 1993). Mental status was retrospectively assessed and four grades were considered: 0 for normal intellectual status, 1 for episodes of hallucinations or confusion, 2 for moderate cognitive impairment and 3 for dementia. In none of the demented PD patients did the neuropsychological pattern evoked AD. Patients with high densities of cortical Lewy bodies or significant vascular changes were excluded. Frequency and intensity of immunodetection of the abnormal Tau triplet were higher in the demented subgroups of PD (grades 2 and 3) than in the non-demented subgroup (grades 0 and 1) in the prefrontal area, temporal cortex and entorhinal cortex but not in either the occipital or cingular cortex (Fig. 1). A quantification of abnormal tau triplet by densitometry showed that, unlike the results obtained in AD patients, the intensity of lesions in the cerebral cortex of most demented PD patients was more severe in the prefrontal area than the temporal area (Fig. 2). This study showed evidence for an Alzheimer-type cortical involvement associated with the cognitive changes in PD. Our results suggest that the unexpected cortical AD lesions, especially those found in the prefrontal cortex, may significantly contribute to the genesis of cognitive changes at least in some PD patients.

The Cortical Pathology of PSP

PSP, also known as Steele-Richardson-Olszewski disease, is a degenerative disorder characterized clinically by supranuclear ophthalmoplegia, pseudobulbar palsy, parkinsonism, axial dystonia and dementia. Neuropathological findings in PSP include NFT and nerve cell loss with gliosis in brainstem, diencephalic and cerebellar nuclei, whereas the cerebral cortex seems to be rarely affected (Jellinger et al. 1980). More recently, a cortical involvement of the degenerating process has been described, with the same features as subcortical NFT but with a different laminar distribution than in AD (Hof et al. 1992b; Hauw et al. 1990).

At the ultrastructural level, the NFT observed in PSP cases are different from those observed in AD. Instead of being composed of PHF, the NFT in PSP are made up of straight filaments (SF) with a diameter of 13 to 22 nm (Tomonaga 1977). Flament et al. (1991) have shown that abnormal Tau species are produced in PSP but they were significantly different from those found in AD.

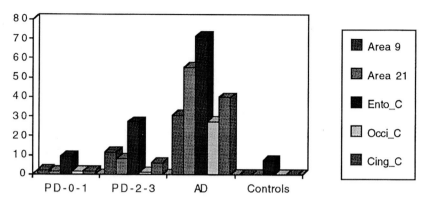

Fig. 2. Quantification of the PTP in the prefrontal BA 9 (Area 9), the temporal BA 21 (area 21), the entorhinal (Ento_C), the occipital (Occi_C) and in the cingular (Cing_C) cortices. The densitometric data (arbitrary values) indicate that the intensity of immunodetection was not uniform among the different subgroups of patients. The intensity was higher in AD and in each group in the entorhinal cortex. The intensity of immunodetection was higher in the demented PD subgroups (PD-2-3) than in the non-demented (PD-0-1; p < .05). By contrast to the results obtained in AD patients, the intensity was higher in the prefrontal than in the temporal cortex of most of the demented PD patients.

Using antibodies against Tau proteins and PHF, we have performed cortical and subcortical biochemical mapping of the neurofibrillary pathology in 4 PSP cases and compared its Tau profiles with those obtained in AD cases. The patients were 67, 71, 69, and 33 years old, respectively, at death. The clinical presentation of cases 1, 2 and 3 were consistent with the diagnosis of PSP and included severe cognitive impairment. Case 4 was a 33-year-old woman with typical features of PSP, but she did not develop dementia.

PTP are distributed in subcortical and cortical areas. The electrophoretic profile of PTP is different from that of AD since a characteristic doublet was found (Tau 64 and 69) instead of the classic Tau triplet (Flament et al. 1991; Vermersch et al. 1994; Buée-Scherrer et al. 1996a; Fig. 1). PTP were detected in almost all subcortical nuclei and brainstem regions of the PSP cases. PTP were also detected in all cortical areas in cases 1 and 2. In case 3, few areas of the parietal, temporal and occipital lobe were spared (Fig. 3). In the young non-demented case 4, abnormal Tau proteins were not detected or were seen in low amounts in several cortical areas (Fig. 4). In the cortical homogenates from PSP cases, the intensities of immunodetection of abnormal Tau proteins were lower than in AD cases. In the demented PSP cases 1, 2 and 3, the intensities were not significantly different between subcortical and cortical regions. Frontal regions were the most affected cortical areas in cases 1, 2 and 3.

Our results show that the neocortical pathology in PSP, as revealed by the presence of PTP, is more extensive than previously thought, but is in good agreement with recent immunohistochemical studies (Hof et al. 1992b; Hauw et al. 1990). The cortical mapping of neurofibrillary pathology in PSP cases exhibits strong differences from that obtained in AD cases. Our results suggest that the

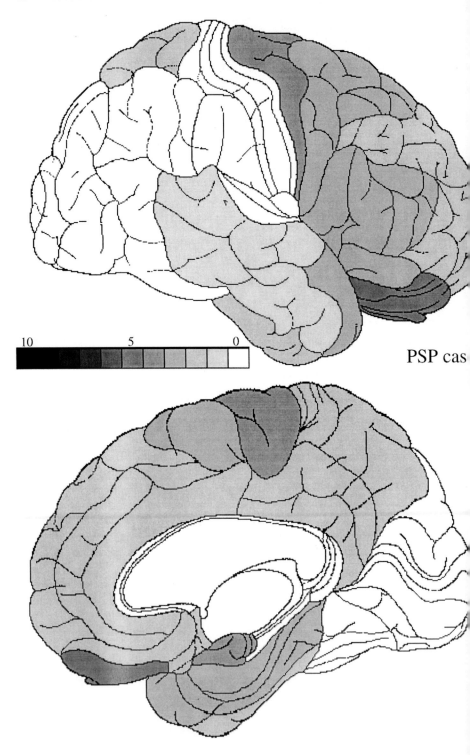

PSP cas

◀ **Fig. 3.** Cortical biochemical mapping of the neurofibrillary degeneration according to the Brodmann classification in PSP case 3 (lateral and medial views). We used a gray scale to display the quantity of PTP in each area from 0 (absence of detection) to 10 (very high concentration of PTP). Note the frontal predominance of involvement.

unexpected cortical lesions in PSP may contribute to the genesis of cognitive changes. Indeed, in the demented PSP cases, cortical lesions are obvious whereas they are not found in the non-demented case. The regional distribution of the PTP may be related to the cognitive impairment.

Conclusion

The biochemical and immunological studies of PTP show that these proteins are excellent markers of neurodegenerative disorders. The different Tau profiles observed in these disorders may be linked to specific vulnerable populations of neurons. Although there are no conclusive data about the correlation between the neurofibrillary pathology and the neuronal loss (Hyman et al. 1995), the immunodetection of PTP is a sensitive and useful method to map and quantify the

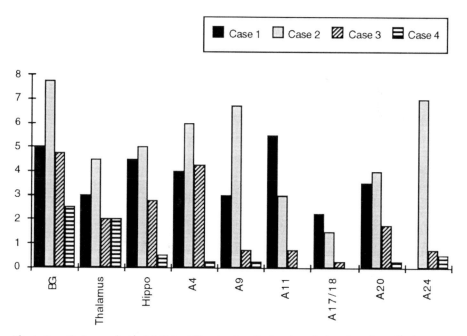

Fig. 4. Quantitative analysis of PTP in different subcortical and cortical regions from the four PSP cases. These proteins were not detected or were in very low amounts in the cortical areas from case 4.

degenerating process. The biochemical quantification probably reflects with good accuracy the quantification of all PHF structures. Whatever the disease, we noted that the presence of PTP in the associative neocortex is invariably linked to a cognitive impairment.

In PSP and in PD, biochemical analyses have shown reduced levels of several neurotransmitters in the cortex, secondary to the alterations of afferent cortical pathways of subcortical origin. The presence of a cortical Tau pathology in demented parkinsonian patients and in PSP demonstrated evidence for cortical neurodegeneration. These data in both disorders call for a reconsideration of the concept of "subcortical dementia."

Acknowledgments

AD2 was developed through a collaboration between UMR-9921 from Montpellier University (Prof. B.Pau, Dr. C.Mourton-Gilles), Sanofi/Diagnostic Pasteur and INSERM.

References

Arriagada PV, Growdon JH, Hedleywhyte ET, Hyman BT (1992) Neurofibrillary tangles but not senile plaques parallel duration and severity of Alzheimer's Disease. Neurology 42: 631–639

Bouras C, Hof PR, Morrison JH (1993) Neurofibrillary tangle densities in the hippocampal formation in a non-demented population define subgroups of patients with differential early pathologic changes. Neurosci Lett 153: 131–135

Braak H, Braak E (1988) Neuropil threads occur in dendrites of tangle-bearing nerve cells. Neuropathol Appl Neurobiol 14: 39–43

Braak H, Braak E (1991) Neuropathological stageing of Alzheimer-related changes. Acta Neuropathol 82: 239–259

Brion JP, Passareiro H, Nunez J, Flament-Durand J (1985) Immunological detection of Tau protein in neurofibrillary tangles of Alzheimer's disease. Arch Biol 95: 229–235

Brion S, Plas J, Jeanneau A (1991) Pick's disease – a clinico-pathological point of view. Rev Neurol 147: 693–704

Buée-Scherrer V, Hof PR, Buée L, Leveugle B, Vermersch P, Perl DP, Olanow CW, Delacourte A (1996a) Hyperphosphorylated Tau proteins differentiate corticobasal degeneration and Pick's disease. Acta Neuropathol 91, 351–359

Buée-Scherrer V, Condamines O, Mourton-Gilles C, Jakes R, Goedert M, Pau B, Delacourte A (1996b) AD2, a phosphorylation-dependent monoclonal antibody directed against Tau proteins found in Alzheimer's disease. Mol Brain Res 39: 79–88

Cummings JL, Benson DF (1984) Subcortical dementia: review of an emerging concept. Arch Neurol 41: 874–879

Delacourte A, Flament S, Dibe EM, Hublau P, Sablonnière B, Hémon B, Sherrer V, Défossez A (1990) Pathological proteins Tau 64 and 69 are specifically expressed in the somatodendritic domain of the degenerating cortical neurons during Alzheimer's disease. Demonstration with a panel of antibodies against Tau proteins. Acta Neuropathol 80: 111–117

Delacourte A, Robitaille Y, Sergeant N, Buée L, Hof PR, Wattez A, Laroche-Cholette A, Mathieu J, Chagnon P, Gauvreau D (1996) Specific pathological Tau protein variants characterize Pick's disease. J Neuropathol Exp Neurol 55: 159–168

Dickson WD, Kress Y, Crowe A, Yen SH (1985) Monoclonal antibodies to Alzheimer neurofibrillary tangles. II: Demonstration of a common antigenic determinant between ANT and neurofibrillary degeneration in progressive supranuclear palsy. Am J Pathol 120: 292–303

Flament S, Delacourte A, Delaère P, Duyckaerts C, Hauw J-J (1990) Correlation between microscopical changes and tau 64 and 69 biochemical detection in senile dementia of the Alzheimer type. Tau 64 and 69 are reliable markers of the neurofibrillary degeneration. Acta Neuropathol 80: 212–215

Flament S, Delacourte A, Verny M, Hauw J-J, Javoy-Agid F (1991) Abnormal tau proteins in progressive supranucelar palsy. Similarities and differences with the neurofibrillary degeneration of the Alzheimer type. Acta Neuropathol 81: 591–596

Giannakopoulos P, Hof PR, Bouras C (1995) Age versus ageing as a cause of dementia. Lancet 346: 1486–1487

Goedert M (1993) Tau protein and the neurofibrillary pathology of Alzheimer's disease. Trends Neurosci 16: 460–465

Gustafson L (1987) Frontal lobe degeneration of non Alzheimer type. II. Clinical picture and differential diagnosis. Arch Gerontol Geriatr 6: 209–223

Grundke-Iqbal I, Iqbal K, Tung YC, Quinlan M, Wisniewski HM, Binder LI (1986) Abnormal phosphorylation of the microtubule-associated protein τ (tau) in Alzheimer cytoskeletal pathology. Proc Natl Acad Sci USA 83: 4913–4917

Hakim AM, Mathieson G (1979) Dementia in Parkinson's disease: a neuropathologic study. Neurology 29: 1204–1214

Hauw J-J, Verny M, Delaère P, Cervera P, He Y, Duyckaerts C (1990) Constant neurofibrillary changes in the neocortex in progressive supranuclear palsy. Basic differences with Alzheimer's disease and aging. Neurosci Lett 119: 182–186

Hof PR, Morrison JH (1990) Quantitative analysis of a vulnerable subset of pyramidal neurons in Alzheimer's disease. II. Primary and secondary visual cortex. J Comp Neurol 301: 55–64

Hof PR, Bierer LM, Perl DP, Delacourte A, Buée L, Bouras C, Morrison JH (1992a) Evidence for early vulnerability of the medial and inferior aspects of the temporal lobe in a 82-year-old patient with preclinical signs of dementia. Regional and laminar distribution of neurofibrillary tangles and senile plaques. Arch Neurol 49: 946–953

Hof PR, Delacourte A, Bouras C (1992b) Distribution of cortical neurofibrillary tangles in progressive supranuclear palsy: a quantitative analysis of six cases. Acta Neuropathol (Berl) 84: 45–51

Hof PR, Bouras C, Perl DP, Morrison JH (1994a) Quantitative neuropathology analysis of Pick's disease cases: cortical distribution of Pick bodies and coexistence with Alzheimer's disease. Acta Neuropathol 87: 115–124

Hof PR, Perl DP, Loerzel J, Steele JC, Morrison JH (1994b) Amyotrophic lateral sclerosis and parkinsonism-dementia of Guam: differences in neurofibrillary tangles distribution and density in the hippocampal formation and neocortex. Brain Res 650: 107–116

Hyman BT, West HL, Gomez-Isla T, Mui S (1995) Quantitative neuropathology in Alzheimer's disease: neuronal loss in high-order association cortex parallels dementia. In: Iqbal K, Mortimer JA, Winblad B, Wisniewski HM (eds) Research advances in Alzheimer's disease and related disorders. Wiley, New York, pp 453–460

Jellinger K, Riederer P, Tomonaga M (1980) Progressive supranuclear palsy: clinico-pathological and biochemical studies. J Neural Trans (suppl 16): 111–128

Lee VMY, Balin BJ, Otvos L, Trojanowski JQ (1991) A68: A major subunit of paired helical filaments and derivatized forms of normal Tau. Science 251: 675–678

Lewis DA, Campbell MJ, Terry RD, Morrison JH (1987) Laminar and regional distribution of neurofibrillary tangles and neuritic plaques in Alzheimer's disease: a quantitative study of visual and auditory cortices. J Neurosci 7: 1799–1808

Matsuo ES, Shin RW, Billingsley ML, Vandevoorde A, Oconnor M, Trojanowski JQ, Lee VMY (1994) Biopsy-derived adult human brain Tau is phosphorylated at many of the same sites as Alzheimer's disease paired helical filament Tau. Neuron 13: 989–1002

Morris JC, McKeel DW, Storandt M, Rubin EH, Price JL, Grant EA, Ball MJ, Berg L (1991) Very mild Alzheimer's disease: informant-based clinical, psychometric, and pathologic distinction from normal aging. Neurology 41: 469–478

Sergeant N, Bussiere T, Vermersch P, Lejeune JP, Delacourte A (1995) Isoelectric point differentiates PHF-Tau from biopsy-derived human brain tau proteins. Neuroreport 6: 2217–2220

Tomlinson BE, Blessed G, Roth M (1968) Observation on the brain of non-demented old people. J Neurol Sci 7: 331–356

Tomonaga M (1977) Ultrastructure of neurofibrillary tangles in progressive supranucelar palsy. Acta Neuropathol 37: 177–181

Vermersch P, Frigard B, Delacourte A (1992 a) Mapping of neurofibrillary degeneration in Alzheimer's disease: evaluation of heterogeneity using the quantification of abnormal Tau proteins. Acta Neuropathol 85: 48–54

Vermersch P, Frigard B, David JP, Fallet-Bianco C, Delacourte A (1992 b) Presence of abnormal tau proteins in the entorhinal cortex in aged non demented subjects. Neurosci Lett 144: 143–146

Vermersch P, Delacourte A, Javoy-Agid F, Hauw J-J, Agid Y (1993) Dementia in Parkinson's disease: biochemical evidence for cortical involvement using the immunodetection of abnormal Tau proteins. Ann Neurol 33: 445–450

Vermersch P, Robitaille Y, Bernier L, Wattez A, Gauvreau D, Delacourte A (1994) Biochemical mapping of neurofibrillary degeneration in a case of progressive supranuclear palsy: evidence of general cortical involvement. Acta Neuropathol 87: 572–577

Vermersch P, David JPh, Frigard B, Fallet-Bianco C, Wattez A, Petit H, Delacourte A (1995 a) Cortical mapping of Alzheimer pathology in brains of aged non-demented subjects. Prog Neuropsychopharmacol Biol Psychiat 19: 1035–1047

Vermersch P, Bordet R, Ledoze F, Ruchoux MM, Chapon F, Thomas P, Destée A, Lechevalier B, Delacourte A (1995 b) Demonstration of a specific profile of pathological Tau proteins in frontotemporal dementia cases. C R Acad Sci 318: 439–445

Neurofibrillary Tangles are Associated with the Differential Loss of Message Expression for Synaptic Proteins in Alzheimer's Disease

*Linda Callahan and Paul D. Coleman**

Summary

Results presented here were obtained by examining the hippocampus of Alzheimer's disease (AD) and control brain by combining three procedures: 1) immunocytochemistry to define neurons containing neurofibrillary tangles (NFT) and neurons free of NFT with 2) *in situ* hybridization for synaptophysin, GAP-43, cathepsin D and total poly A + RNA. 3) Sections were counter stained with cresyl violet. Results demonstrated several-fold loss of message for synaptophysin and GAP-43 in neurons containing NFT relative to message levels detected in neighboring neurons that did not contain NFT. Neurons from AD brain that did not contain NFT expressed synaptophysin message at a level similar to that seen in neurons from control brain. The decreased level of expression for synaptophysin and GAP-43 message was selective, since expression of total poly A + message was only slightly reduced in neurons containing NFT, and since expression of message for the lysosomal enzyme cathepsin D increased in neurons containing NFT. These data introduce the concept of changing profiles of expression within single neurons as the disease progresses in each neuron.

Introduction

The bulk of current evidence suggests that the two neurobiological parameters that best correlate with the degree of dementia in Alzheimer's disease (AD) are increased density of neurofibrillary tangles (NFT) and decreased density of synapses (e.g., Arriagada et al. 1992; DeKosky and Scheff 1990; Terry et al. 1991). A potential relationship of synaptic loss to degree of dementia is intuitively appealing since synapses are the morphological substrate by which neurons connect with each other to transmit, process and store information, all abilities which are severely decreased in AD. The hypothesis we propose derives heavily from the fact that microscopic examination of AD brain that has been prepared to reveal

* Dept Neurobiology & Anatomy, University Rochester Medical Center, 601 Elmwood Avenue, Box 603, Rochester, NY 14642, USA

B. T. Hyman / C. Duyckaerts / Y. Christen (Eds.)
Connections, Cognition, and Alzheimer's Disease
© Springer-Verlag Berlin Heidelberg 1997

NFT shows that neurons free of NFT may be close neighbors of neurons that contain NFT. Our hypothesis is that it is the NFT-containing neurons that are (largely) responsible for the loss of synapses that has been demonstrated in the AD brain.

Since NFT are found in the cell body and the majority of synapses made by a cell are on distal axon terminals, coupling a characterized cell body with the synapses it forms is a formidable task. However, since quantification of message for synapse-associated proteins has been shown to be a reflection of that synaptic protein (Eastwood et al. 1994) it seemed that message in the cell body for a synapse-associated protein would provide a reasonable estimate of the ability of that cell body to provide the protein building blocks of the synapse. We therefore undertook a test of our hypothesis by combining immunocytochemistry (ICC) to identify NFT-bearing and NFT-free neurons with *in situ* hybridization to assess the level of message for the synapse-associated protein, synaptophysin.

Results and Discussion

Sections of the hippocampal CA1 region of AD brain show clumps of abundant grain density, representing synaptophysin message over immuno-negative (NFT-free) neurons. Closely neighboring neurons in the same section that were immuno-positive for monoclonal antibody 69 (a gift of Dr. S.-H. Yen), representing neurons containing paired helical filaments, showed a greatly reduced grain density, signifying reduced message level for synaptophysin. Figure 1 illustrates this material and provides an indication of the degree of reduction of synaptophysin message in tangle-bearing neurons.

Expression of synaptophysin message in non-tangle-bearing neurons in AD brain is at a level similar to that seen in neurons from control brain mounted side-by-side on the same slide as sections of AD brain. The average grain count over CA1 neurons from control brain was 250 ± 12 (s.e.m.). The average grain count over tangle-free neurons from AD brain was 228 ± 14. This 9 % reduction over tangle-free neurons from AD brain was not statistically significant ($t = 1.24$; $p > 0.05$). Thus, it appears that even in brain regions that are severely affected by AD, neurons remain that are expressing message for this synaptic marker at levels equivalent to the level seen in control brain. Presumably, these neurons in AD brain represent a residual capacity.

Earlier work from our laboratory has also shown an approximately three-fold reduction in level of message for the axonal growth-associated protein, GAP-43, in tangle-bearing neurons relative to neighboring tangle-free neurons in AD brain (Callahan et al. 1994). However, these reductions of message for synaptophysin and GAP-43 appear to be selective, rather than a consequence of a generalized reduction in message expression in neurons that have developed NFT. A previous quantitative *in situ* hybridization study by Griffin et al. (1990) demonstrated an approximately 20 % reduction in poly A + RNA in tangle-bearing neurons in AD brain, a reduction that is far less that the several-fold reduction in

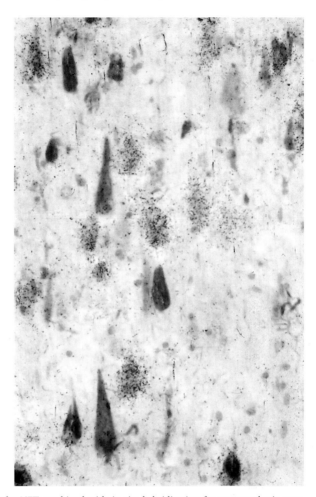

Fig. 1. Immunocytochemistry for NFT combined with *in situ* hybridization for synaptophysin message. Cresyl violet counterstain. The synaptophysin riboprobe was derived from a clone containing the entire coding region. For *in situ* hybridization this was hydrolized to an average size of approximately 200 bases. Monoclonal antibody 69 to a conformational epitope of paired helical filaments reveals both intracellular and extracellular "tombstone" tangles. Tombstone tangles are eliminated from consideration by requiring clear evidence of a nucleus within each cell analyzed. Note greatly reduced density of grains over neurons containing NFT in comparison to adjacent neurons without NFT. Also note neuron in the upper right without NFT but also without appreciable synaptophysin message expression. Such neurons are rare, but their characterization may give clues to the progression of disease at a cellular level. 40X objective.

synaptophysin and GAP-43 message that we find in neurons containing NFT. In our hands, a poly U probe used to reveal total messenger RNA produced results that are in essential agreement with those of Griffin et al. (data not shown). These data suggest that near maintenance of total message level in many tangle-bearing neurons in the face of major losses of message for synapse and growth-associated proteins probably requires that expression of some messages be up-regulated at some stage of the formation of NFT.

Previous work by Nixon and his collaborators (e.g., Cataldo et al. 1995) demonstrated an increased expression of selected lysosomal enzymes, including cathepsin D, by neurons in the AD brain. Combining immunocytochemistry with mAb 69 to reveal paired helical filaments with *in situ* hybridization for cathepsin D message reveals that message for this lysosomal enzyme is increased in neurons containing NFT (Fig. 2).

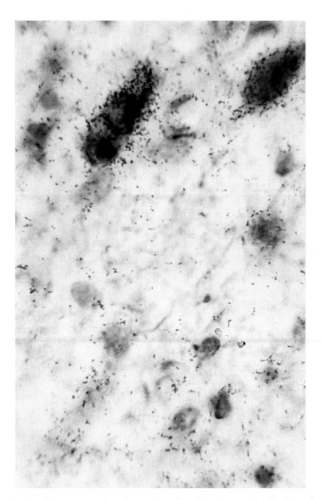

Fig. 2. Immunocytochemistry for NFT as in Figure 1, with *in situ* hybridization for cathepsin D and counterstaining with cresyl violet. Cathepsin D probe was obtained from a full length clone, a kind gift of S. Kornfeld. Probe was hydrolized to an average size of approximately 200 bases. Note heavy grain density over tangle-bearing neurons, with relatively low grain density over faintly visible, cresyl violet stained, tangle-free neurons.

Thus, our work to date has indicated that the expression profile is altered in neurons when they develop NFT. The changes appear to be such that we might generalize that expression of messages associated with the synapse and with growth of processes may be reduced, whereas expression of messages associated with stress responses of the neuron may be increased as the NFT develops. Expression of other classes of messages must also change. Certainly as the disease progresses toward the death of any given single neuron, the profile of messages expressed within that neuron must undergo a number of transitions from the expression profile of a healthy neuron to the expression profile of a neuron about to die and, finally, to total loss of expression.

These data demonstrating selective loss of message for a synapse-associated protein in neurons bearing NFT demonstrate a central role for NFT in the pathophysiology of AD and place the NFT, or events closely associated with the NFT, as a major factor in the disconnections and cognitive declines of Alzheimer's disease.

Acknowledgments

The work reported here was supported by grants from the National Institute on Aging, The Markey Foundation and The American Health Assistance Foundation.

References

Arriagada PV, Growdon JH, Hedley-Whyte ET, Hyman BT (1992) Neurofibrillary tangles but not senile plaques parallel duration and severity of Alzheimer's disease. Neurology 42: 631–639

Callahan LM, Coleman PD (1995) Neurons bearing neurofibrillary tangles are responsible for selected synaptic deficits in Alzheimer's disease. Neurobiol Aging 16: 311–314

Callahan LM, Selski DJ, Martzen MR, Cheetham JE, Coleman PD (1994) Preliminary evidence: decreased GAP-43 message in tangle-bearing neurons relative to adjacent tangle-free neurons in Alzheimer's disease parahippocampal gyrus. Neurobiol Aging 15: 381–386

Cataldo AM, Barnett JL, Berman SA, Li J, Quarless S, Bursztajn S, Lippa C, Nixon RA (1995) Gene expression and cellular content of cathepsin D in Alzheimer's disease brain: evidence for early up-regulation of the endosomal-lysosomal system. Neuron 14: 671–680

DeKosky ST, Scheff SW (1990) Synapse loss in frontal cortex biopsies in Alzheimer's disease: correlation with cognitive severity. Ann Neurol 27: 457–464

Eastwood SL, Burnet PW, McDonald B, Clinton J, Harrison PJ (1994) Synaptophysin gene expression in human brain: a quantitative in situ hybridization and immunocytochemical study. Neuroscience 59: 881–892

Griffin WS, Ling C, White CL, Morrison-Bogorad M (1990) Polyadenylated messenger RNA in paired helical filament-immunoreactive neurons in Alzheimer disease. Alzheimer Disease Assoc Disorders 4: 69–78

Terry RD, Masliah E, Salmon DP, Butters N, DeTeresa R, Hill R, Hansen LA, Katzman R (1991) Physical basis of cognitive alterations in Alzheimer's disease: synapse loss is the major correlate of cognitive impairment. Ann Neurol 30: 572–580

Morphologic and Neurochemical Characteristics of Corticocortical Projections: Emergence of Circuit-specific Features and Relationships to Degenerative Changes in Alzheimer's Disease

P. R. Hof*, E. A. Nimchinsky, L. G. Ungerleider, J. H. Morrison

Summary

Detailed regional and laminar analyses of the neuropathological lesions in Alzheimer's disease (AD) indicate that subsets of neocortical and hippocampal neurons are compromised, leading to the disruption of corticocortical pathways which in turn may subserve the dementia observed clinically. Previous analyses of the morphologic and biochemical characteristics of the neurons that are vulnerable to degeneration and neurofibrillary tangle formation in AD have demonstrated that a subpopulation of pyramidal neurons containing high concentration of neurofilament protein in layers III and V of neocortical association areas is dramatically affected in AD, whereas these neurons are relatively spared in primary sensory and motor cortices. In addition, in normal brains, these cells exhibit highly specific regional and laminar distribution patterns in the neocortex and their morphology and location suggest that they are the cells of origin of long corticocortical projections. However, the degree to which these neurons contribute of different long cortical pathways in the primate brain is not known. To explore this issue, we performed an extensive quantitative analysis of more than 30 corticocortical pathways in the monkey neocortex, combining tract-tracing and immunohistochemistry. These studies revealed that the proportion of neurofilament protein-containing neurons varies substantially among corticocortical projections, with low numbers of immunoreactive neurons in the limbic and short corticocortical connections (0–30%), intermediate numbers in callosal connections (30–45%), and very high numbers (60–100%) in long association pathways linking polymodal regions. In the visual system, there are higher proportions of immunoreactive neurons in the occipitoparietal pathway (up to 85%) compared to the occipitotemporal pathway (up to 40%). Interestingly, a similar pattern is observed in the visual pathways for the kainate receptor subunit proteins GluR5/6/7, whereas the AMPA GluR2/4 and NMDAR1 subunits are ubiquitous, indicating a possible link between excitotoxic mechanisms and the presence of neurofilament protein and specific combination of glutamate receptor subunits in certain populations in AD. In addition, combined cell loading,

* Fishberg Research Center for Neurobiology, Box 1065, Mount Sinai School of Medicine, One Gustave L. Levy Place, New York, NY 10029, USA

B. T. Hyman / C. Duyckaerts / Y. Christen (Eds.)
Connections, Cognition, and Alzheimer's Disease
© Springer-Verlag Berlin Heidelberg 1997

tract-tracing and immunohistochemistry analyses demonstrated that there is a considerable degree of regional and laminar variability among the different corticocortical projections, not only in terms of neurofilament protein content but also regarding cellular morphology. Several studies have shown that neurofilament protein is involved in the formation of neuropathological lesions in several neurodegenerative disorders. Although neurofilament protein is a useful and reliable marker of neuronal vulnerability in AD, the differential distribution of neuropathologic lesions in dementing illnesses and the apparent discrepancies observed in neurofilament protein and glutamate receptor subunit proteins expression among corticocortical projections suggest that the phenotype of the affected cells in neurodegenerative disorders is related not solely to the presence of certain molecules but to morphological features and connectivity constraints as well. Comparison of results from the macaque monkey analyses with quantitative neuropathological data in human will be important to characterize further the degree to which the vulnerable neurons in AD are homologous to the neurons furnishing corticocortical connections in nonhuman primates. Such comparative studies may also provide valuable information on the involvement of different sets of neurochemically identifiable neuronal populations in the course of various neurodegenerative illnesses.

Introduction

The key pathogenetic events that lead to dementia are not yet fully understood, although numerous hypotheses regarding the formation of the neuropathological lesions observed in AD have been proposed. The distribution of pathological changes in AD brains suggests that structurally and functionally AD is predominantly a disease of the cerebral cortex. However, it cannot be considered a global loss of cortical function since dementia involves certain populations of neurons displaying a specific regional and laminar distribution and connectivity patterns, whereas other neuron types are spared (for review, see Morrison 1993; Hof and Morrison 1994). It is therefore likely that the neuronal vulnerability in AD can be related to morphologic and biochemical characteristics of identifiable neuronal populations and connections.

Neuroanatomical analyses of the non-human primate brain is central to our understanding of cortical cellular organization and circuitry, and in the past decades many such studies have been crucial in regard to the organization of corticocortical circuits and interactions between neocortex and the hippocampal formation in primates, including human. In addition, the use of specific histochemical techniques has allowed for the investigation of issues such as neuron typology, connectivity and circuit distribution within the context of neurochemical identity, and for the development of quantitative organizational schemes of the cerebral cortex that link neurochemical markers to specific groups of neurons or circuits as well. In view of the recent increase in functional and chemically specific studies of the primate brain, it is possible that such organizational models can eventually be extended to the human cerebral cortex, in spite of the obvious limitations as to the application of experimental procedures to human brain materials. However, many of the histochemical procedures work reliably in the postmortem human brain and, to some extent, correlations across species can be drawn. These correlations have been particularly useful in regard to issues pertaining to the neuropathology of AD as well as to the normal functional anatomy of the human cerebral cortex, although directly transposing information inferred from analyses of the monkey to the human brain should always be considered with caution as obvious differences exist in cortical organization among primates. Clearly, recent advanced in functional brain imaging have also greatly contributed to our understanding of human functional neuroanatomy in normal conditions as well as in aging and dementing illnesses (Haxby et al. 1991; Grady et al. 1994; Ungerleider 1995; DeYoe et al. 1996).

In the present article, we review data that suggest the existence of relationships between the distribution of pathologic changes in AD and the localization of specific elements of the cortical circuitry that are affected by these alterations. Based on extensive analyses of the neurochemical coding of macaque monkey cortical connectivity, we describe observations that relate the biochemical phenotype of a given neuron to its possible role in corticocortical circuitry and discuss potential molecular and anatomic elements participating in a general profile of cellular vulnerability or resistance to the degenerative process in AD.

Distribution of Neuropathologic Lesions in AD

Quantitative analyses of neuronal loss and the distribution of neurofibrillary tangles (NFT) and senile plaques (SP) in neocortex and hippocampal formation have revealed that there are links between the localization of these lesions to specific cortical regions and layers, and the cells of origin of long corticocortical and hippocampal projection (Pearson et al. 1985; Rogers and Morrison 1985; Duyckaerts et al. 1986; Lewis et al. 1987; Morrison et al. 1987; Braak et al. 1989; Hof and Morrison 1990, 1994; Hof et al. 1990a; Morrison 1993). High densities of NFT in layers II and V of the entorhinal cortex are consistently found in AD cases, whereas NFT are observed primarily within layers III and V in the neocortex. Primary sensory and motor areas have far fewer NFT than association cortices, such that there is an approximately 20-fold increase in the number of NFT in the secondary visual cortex compared to primary visual cortex, and a further doubling in NFT density in visual association areas in the inferior temporal cortex (Pearson et al. 1985; Lewis et al. 1987; Hof and Morrison 1990; Arnold et al. 1991). Also, there are regional differences in laminar NFT distribution among different neocortical areas. For example, most of the NFT are in layer III in the primary and secondary visual areas, whereas there is a shift toward layer V in the inferior temporal cortex. Senile plaques are generally more numerous in the neocortex, where they predominate in the superficial layers, that in the hippocampal formation (Lewis et al. 1987; Arnold et al. 1991). Thus, distribution and density of these lesions exhibit certain organizational patterns that are useful in linking alterations of the cortical circuitry to clinical symptomatology.

The use of specific immunohistochemical and tract-tracing techniques in nonhuman primates has considerably expanded our knowledge of neuronal typology and connectivity in relation to neurochemical phenotype. The regional and laminar specificity of lesion distribution in the AD neocortex described above suggests the existence of a close association of NFT with the cells of origin of corticocortical projections, whereas the distribution of SP appears to correlate to some degree with the termination of the corticocortical projections (Pearson et al. 1985; Rogers and Morrison 1985; Duyckaerts et al. 1986; Lewis et al. 1987; Braak et al. 1989; Hof et al. 1990a; Hof and Morrison 1990). Studies in the nonhuman primate have shown that corticocortical projections can be classified as feedforward, feedback, and lateral connections (Felleman and Van Essen 1991; Hof and Morrison 1995). Feedforward connections ascend within a hierarchically organized system (i.e., from a primary sensory area to an association area), whereas feedback projections descend the same hierarchy. Lateral projections connect cortical regions that are at the same level in a given hierarchy. Feedforward connections originate from neurons located in the superficial layers and terminate in layer IV and the deep portion of layer III of the target cortical region. Feedback projections emanate from the deep layers and terminate in layers I and VI. Lateral connections arise from the deep layers and project to all layers of the target regions (Felleman and Van Essen 1991; Hof and Morrison 1995). The localization of NFT in layers III and V suggests that all of these projections are potentially affected in AD. The fact that in association cortex layer V contains generally higher NFT densities than the superficial layers suggests that feedback as well as lateral projections are affected in AD. The feedforward projections are clearly damaged in the primary and secondary sensory cortices, since in these areas, NFT are concentrated in layer III, where most of the efferent corticocortical neurons are located (Lewis et al. 1987). Well-defined projections in the hippocampal formation also display patterns of lesion distribution comparable to those seen in neocortical circuits (Hyman et al. 1984, 1986, 1990; Braak and Braak 1991; Senut et al. 1991). For example, the perforant pathway is severely and consistently affected in AD, and the presence of NFT in its neurons of origin in layer II of the entorhinal cortex is correlated with a high density of SP in its termination zone in the molecular layer of the dentate gyrus (Hyman et al. 1984, 1986; Senut et al. 1991).

These observations suggest that the pathologic alterations visible at a regional level in the cerebral cortex reflect the disconnection of identifiable cortical pathways and the differential involvement of identifiable neuronal subsets (Rogers and Morrison 1985; Lewis et al. 1987; Morrison et al. 1987; Hof et al. 1990a; Hof and Morrison 1990; Hyman et al. 1990). This view is also supported by previous observations of AD cases with atypical clinical presentation. Most of these cases present with visual symptomatology as the first clinical evidence of the degenerative process, preceding the classical memory impairment and cognitive deficits of AD (Benson et al. 1988). In these cases, computer assisted tomography and magnetic resonance imaging, or pathologic examination at autopsy, demonstrated a substantial atrophy of the parieto-occipital areas, referred to as posterior cortical atrophy (Benson et al. 1988; Hof et al. 1989, 1990b, 1993a; Graff-Radford et al. 1993; Levine et al. 1993). We have reported a series of such cases, where a complex visual deficit in visuospatial skills and motion detection is superimposed on the dementia and memory defects typically seen in AD (Hof et al. 1989, 1990b, 1993a). Neuropathologically, in all of these cases the primary (Brodmann's area 17) and secondary visual cortices (Brodmann's areas 18 and 19) always had strikingly higher NFT counts compared to the classic AD pattern, whereas in the prefrontal cortex, NFT counts were always lower than in generally seen in AD. Senile plaques were more numerous in the occipital areas than usually seen in AD and the visual association cortices of the inferior parietal lobe (area 7b) and the posterior cingulate cortex (area 23) contained more SP and NFT in the AD cases with Bálint's syndrome. Thus, a caudal shift in the distribution of pathologic changes had occurred in these cases, as well as an increase in lesion counts in the inferior parietal, primary, and secondary visual cortices, suggesting that the occipitoparietal set of connections that subserves visuospatial analysis appears to be more affected in these cases than is usually the case in AD.

Importantly, the lesion distribution pattern seen in AD appears to be quite specific compared to other dementing disorders. In this context, the NFT distribution in amyotrophic lateral sclerosis/parkinsonism-dementia complex of Guam suggest that other classes of pyramidal neurons are involved than in AD, since NFT predominate in layers II and III in all neocortical areas investigated (Hof et al. 1991a). Comparably in Pick's disease, Pick bodies are preferentially located in layers II and VI (Hof et al. 1994). Interestingly, neurons in layers II, superficial III, and VI are known to provide short-distance corticocortical connections (Jones and Wise 1977; Felleman and Van Essen 1991; Hof et al. 1995a). The regional and laminar patterns of distribution may thus be specific for a particular form of dementing illness, and therefore dementias could be understood as complex syndromes with differential involvement of many elements of the cerebral circuits, which could lead to the clinical deficits defining each disorder.

Characteristics of the Neurons at Risk in AD

Not all neuron types are prone to NFT formation in AD. Large pyramidal cells in layers III and V of neocortex represent the most affected cell class, and these neurons are efferent cells that send long projections to other cortical sites as well as subcortical structures. The spiny stellate cells and small pyramidal cells in layer IV are remarkably resistant, as are many inhibitory interneuron classes. Similarly, most of the large pyramidal efferent neurons in layers II and V of the entorhinal cortex and in the CA1 field and subiculum contain NFT in AD. Thus, the notion of differential vulnerability relies on a detailed and cohesive definition of neuronal classes that includes several interactive criteria such as morphology, regional and laminar location, connectivity, and neurochemical phenotype (Morrison 1993; Hof et al. 1994).

Subgroups of pyramidal neurons in the cerebral cortex of both human and monkey have been shown to be enriched in neurofilament protein (Campbell and Morrison 1989; Hof and Morrison 1995). Interestingly, neurofilament as well as several other cytoskeletal proteins have been implicated in NFT formation (Trojanowski et al. 1993). Extensive regional heterogeneity exists in the size, density and laminar distribution of neurofilament protein-containing cells in frontal, cingulate, temporal and occipital areas of both monkey and human cortex (Campbell and Morrison 1989; Hof et al. 1900a, 1995b; Hof and Morrison 1990, 1995; Nimchinsky et al. 1995). As discussed below, the distribution of neurofilament protein-containing cells corresponds to the distribution of corticocortically projecting cells, as demonstrated in transport studies in monkey cortex (Hof et al. 1995a). The laminar distribution of neurofilament protein-containing neurons in visual areas of normal human cortex, as well as superior frontal and inferior temporal cortices, is comparable to that of NFT, and the layers displaying high NFT densities suffer severe loss of neurofilament-containing neurons in AD (Hof et al. 1990a; Hof and Morrison 1990). Similarly, projection neurons in layers II and V of the entorhinal cortex and in the subiculum have a very density of neurofilament protein in the healthy human brain, and these regions display a dramatic loss of neurofilament protein-containing neurons in AD (Morrison et al. 1987; Vickers et al. 1992). Quantitative analyses in AD have demonstrated an extensive loss of neurofilament protein-containing neurons in layers III and V of both the inferior temporal and superior frontal cortex, the largest neurons being the most affected (up to 90 % loss in this particular cell class; Fig. 1A, B; Hof et al. 1990a). However, a much lower degree of neurofilament protein-containing neuron loss is observed in areas 17 and 18, with cell loss restricted to layer IVB and to the Meynert cell class (about 25 % loss) in area 17, and limited to deep layer III (about 18 % loss) in area 18, which parallels the lower incidence of NFT in these regions and the fact that affected neurons are precisely those forming long efferent connections from these areas (Hof and Morrison 1990).

Further evidence supports the observation that the loss of neurofilament protein-containing neurons is correlated with NFT densities. In elderly nondemented cases, layer II of the entorhinal cortex contains neurofilament protein-immunoreactive neurons that also are immunoreactive to tau protein and thioflavine S-positive materials (Fig. 1C–F), suggesting the presence of transitional, early forms of NFT, whereas in the frontal cortex there are very few such NFT and the majority of neurofilament protein-containing neurons in layers III and V appear intact at this stage (Vickers et al. 1992). In severe AD cases, early forms of NFT are seen in the frontal cortex, whereas NFT in the entorhinal cortex have lost their immunoreactivity to tau and neurofilament proteins and are stained only with thioflavine S (representing for the most part extracellular NFT), indicating that NFT formation undergoes a dynamic process in affected neurofilament protein-containing neurons (Fig. 1C-F; Vickers et al. 1992, 1994a). These data suggest that the presence of high perikaryal and dendritic concentrations of neu-

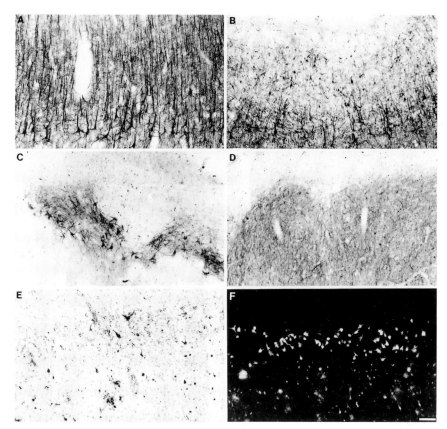

Fig. 1. Neurofilament protein-containing neurons and NFT in the superior frontal cortex (area 9, **A, B**) and layer II of the entorhinal cortex (**C–F**), in a non-demented elderly case (**A, C, E**) and AD case (**B, D, F**). Tissues were stained for neurofilament protein (**A-D**), hyperphosphorylated microtubule-associated protein tau (**E**), or with thioflavine S (**F**). Note the loss of neurofilament protein-containing neurons in both regions in the AD case (compare **A, C** to **B, D**). There are a few tau protein-positive NFT in the entorhinal cortex of the control case (**E**), whereas most affected neurons are end-stage NFT in the AC case (**F**). Scale bar (on **F**) = 100 μm

rofilament protein is probably one of the key neurochemical characteristics of the neurons that are vulnerable in AD. Interestingly, cellular levels of neurofilament protein change during human brain aging in certain neurons. Pyramidal neurons in the CA 1 field of the hippocampus have very low somatic and intradendritic neurofilament protein levels in young adults (Vickers et al. 1994a), but elderly individuals demonstrate a substantial increase in neurofilament protein levels that correlates with the high vulnerability of these neurons to NFT formation in the aged brain, indicating that the neurochemical profile of neurons can change in an age-related manner. In addition, immunoblotting analyses in AD cases have shown that there is an increased amount of neurofilament protein compared to control cases (Vickers et al. 1994a), suggesting that changes in neurofilament protein levels are further potentiated in the course of dementia. Thus, age-related factors involving modifications of the cytoskeleton and leading to the formation of pathologic proteins may represent a fundamental step in the development of NFT in the aging brain, which can be further potentiated in dementing disorders.

In contrast to these pyramidal neurons, distinct classes of interneurons containing the calcium-binding proteins parvalbumin, calbindin and calretinin are in most cases resistant to the degenerative process, even in severe AD cases with high densities of NFT and SP (Hof and Morrison 1991; Hof et al. 1991 b, 1993 a). Parvalbumin and calretinin-immunoreactive neurons are generally not vulnerable and show a well-preserved morphology and staining pattern in AD (Hof et al. 1991 b, 1993 b). Calbindin-containing interneurons in layers II and III are resistant to degeneration (Hof and Morrison 1991), whereas a smaller population of calbindin-immunoreactive cells in layer V is affected in AD cases with high NFT densities. The differential vulnerability could reflect distinct connectivity patterns of various pools of calbindin-containing neurons, as well as specific interactions with their target cells.

Neurofilament Protein as a Marker of Corticocortical Pathways in the Primate Neocortex

The neurochemical characteristics of the neuronal subsets that furnish different types of corticocortical connections have been only partially determined. In recent years, several cytoskeletal proteins have emerged as reliable markers to distinguish subsets of pyramidal neurons in the cerebral cortex of primates. In particular, neurofilament protein has revealed a consistent degree of regional and laminar specificity in the distribution of a subpopulation of pyramidal cells in the primate cerebral cortex (Figs. 2–3; Campbell and Morrison 1989; Hof et al. 1990 a, 1995 b; Hof and Morrison 1990, 1995; Nimchinsky et al. 1995). To characterize further the properties of neurofilament protein in cortical circuitry and its potential role in AD, it is useful to correlate the regional and laminar distribution of neurons containing high levels of this marker with the distribution of corticocortical pathways. Also, it is necessary to explore in detail the possible relationships between the morphologic features of identified corticocortically projecting neurons and the presence of neurofilament protein in the somatodendritic domain of such cells.

In this context, the visual system of primates is particularly well understood in terms of functional anatomical organization (Mishkin et al. 1983; Van Essen

---

Fig. 2. Neurofilament protein-immunoreactive neurons in orbitofrontal cortex (area 13, **A**), the ▶ mid portion of the insula (**B**), the anterior cingulate cortex (area 24a, **C**), and the posterior cingulate cortex (area 23b, **D**) in the macaque monkey. Note the regional and striking laminar differences in neuronal densities among these areas. Scale bar (on **D**) = 100 μm. (Adapted from Hof et al. 1995).

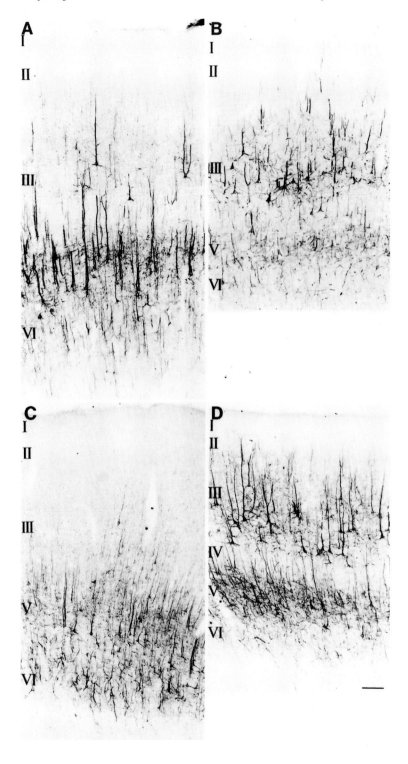

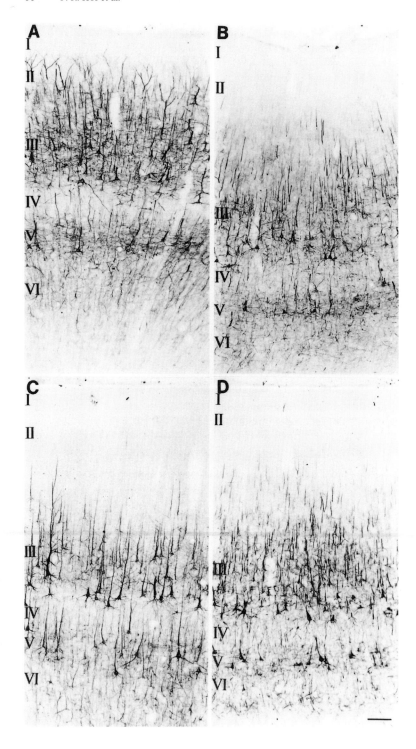

◀ **Fig. 3.** Neurofilament protein-immunoreactive neurons reciprocally connected to cortical regions in the principal sulcus (area 46, **A**), the superior temporal sulcus (area STP, **B**), and the parietal cortex (area 7a, **C**; area LIP, **D**) in the macaque monkey. Note the laminar and regional differences in neuronal densities. The labeled cells predominate in the deep portion of layer III and in layer V. Compare with limbic system-associated areas on Figure 2. Scale bar (on **D**) = 100 μm. (Adapted from Hof et al. 1995)

and Maunsell 1983; DeYoe and Van Essen 1988; Livingstone and Hubel 1988; Clarke and Miklossy 1990; Felleman and Van Essen 1991; Haxby et al. 1991; Watson et al. 1993; Grady et al. 1994; Ungerleider 1995; DeYoe et al. 1996). Visual function is subserved at the cortical level by a large number of areas with specific physiologic properties and connectivity patterns. However, precise indicators of the degree of anatomic specialization has not yet been defined for many of these cortical regions. In a recent analysis of the macaque monkey visual cortex, we provided a precise description of neurofilament protein immunoreactivity to define quantitatively the regional boundaries of the various visual cortical areas (Hof and Morrison 1995). The analysis of the distribution and density of neurofilament protein-containing neurons in the macaque monkey visual cortex reveals that as many as 28 areas can be distinguished based to immunolabeling criteria, with each region displaying a characteristic and homogeneous distribution of labeled neurons (Fig. 4). Certain areas could be grouped according to spe-

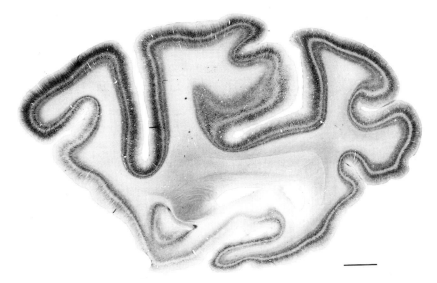

Fig. 4. Low power photomontage of a coronal section through the occipito-parieto-temporal junction of a macaque monkey, stained for neurofilament protein immunoreactivity. This section contains many of the mediotemporal visual association areas. Note the differences in regional and laminar distribution of immunoreactive neuron in the cerebral cortex. Scale bar = 1.5 mm. (Adapted from Hof and Morrison 1995)

cific staining patterns or density trends. For instance, within the occipitotemporal (parvocellular) pathway, areas V3, V4, and regions within the inferior temporal cortex are characterized by a distinct population of neurofilament-rich neurons in layers II–IIIa, whereas areas located in the parietal cortex and part of the occipitoparietal (magnocellular) pathway have a consistent population of large labeled neurons in layer V, which are not observed in "parvocellular" areas (Hof and Morrison 1995). The mediotemporal areas MT and MST display a unique population of densely labeled, highly polymorphic neurons in layer VI. In addition, quantitative analysis of the laminar distribution of the labeled neurons demonstrates that the visual cortical areas can be further divided into four hierarchical levels based upon the ratio of neuron counts between infragranular and supragranular layers, with the first (areas V1, V2, V3, and V3A) and third (temporal and parietal regions) levels characterized by low rations and the second (areas MT, MST, and V4) and fourth (frontal regions) levels characterized by high to very high ratios (Fig. 5). Such density trends may correspond to differential representation of corticocortically and cortico-subcortically projecting neurons at several functional steps in the integration of the visual stimuli. Interestingly, progressive increases in cell density up to 2.4-fold in layer IVB and 4.1-fold for the Meynert cells in area V1 correspond with the degree of eccentricity in the visual field representation, demonstrating another possible correlation

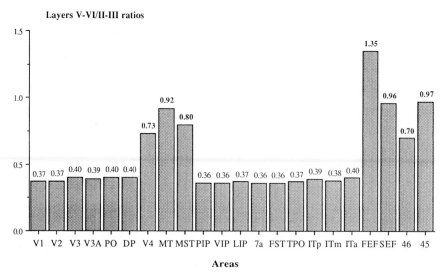

Fig. 5. Mean rations of neurofilament protein-immunoreactive neuron counts between the Layers V–VI and II–III in a series of visual cortical areas. Note that areas V4, MT, MST, 46, 45 and the FEF, SEF had significantly higher ratios due to increased densities of labeled neurons in layers V and VI compared to all of the other cortical fields (bold values, p <0.001). Such differences may reflect differential connectivity patterns of these areas at different levels in the cortical hierarchy. Regions are labeled according to Felleman and Van Essen (1991) and Hof and Morrison (1995). (Adapted from Hof and Morrison 1995)

between distribution of neurofilament protein and that of neurons with identifiable morphology and known connections (Hof and Morrison 1995).

The density of neurofilament protein-immunoreactive neurons has also been shown to vary across corticocortical pathways in macaque monkeys (Campbell et al. 1991; Hof and Morrison 1995; Hof et al. 1995a, 1996). We have used the this marker to examine and quantify the distribution of a subset of corticocortically projecting neurons in a number of long ipsilateral corticocortical pathways in comparison to short corticocortical, commissural, and limbic connections. The results of tract-tracing experiments combined with immunohistochemistry demonstrate that long association pathways interconnecting the frontal, parietal and temporal neocortex have a high representation of neurofilament protein-enriched pyramidal neurons (45–90 % of the retrogradely labeled cells also contained neurofilament protein), whereas short corticocortical, callosal, and limbic pathways are characterized by much lower numbers of such neurons (4–35 % of double labeling; Table 1; Hof et al. 1995a). These data suggest that different types of corticocortical connections have differential representation of highly specific neuronal subsets that share common neurochemical characteristics, thereby determining regional and laminar cortical patterns of morphological and molecular heterogeneity. These differences in neuronal neurochemical phenotype among corticocortical circuits may have considerable influence on cortical processing, and may be directly related to the type of integrative function subserved by each cortical pathway. With respect to the dramatic involvement of neurofilament protein-containing neurons in AD, these data support the hypothesis that

Table 1. Percentage of neurofilament protein-containing neurons in corticocortical projections connecting the frontal, parietal, temporal, and cingulate cortices[a]

Areas	46	7a	STP	24	23
46 ipsi	16/12	42/37	75/72	11/29	33/31
46 contra	18/23	–	–	–	–
7a ipsi	46/46	28/22	47/44	33/26	42/48
7a contra	–	41/32	–	–	–
STP ipsi	84/84	81/78	17/20	–	–
STP contra	–	–	34/39	–	–
24 ipsi	8/15	–	10/13	6/23	6/21
24 contra	5/13	–	–	–	–
23 ipsi	–	18/19	–	18/23	23/19
23 contra	–	21/23	–	–	–
11/13	–	–	24/22	–	–
Insula	–	–	21/26	–	–

[a] Data represent percentages and were pooled from six macaque monkeys. For each projection the first value is from layers II–III and the second value from layers V–VI. A missing value means that the projection was not quantified in our study or that too few neurons were observed to establish a reliable data set. Note the high percentages among connections linking the frontal (46), temporal (STP) and parietal (7a) regions, whereas the commissural, limbic and short association connections display much lower values

neurofilament protein may be crucially related to the development of selective neuronal vulnerability and subsequent disruption of corticocortical pathways that may lead to the severe impairment of cognitive function observed in senile dementia.

In a related set of experiments we analyzed the neurochemical phenotype of neurons furnishing feedforward and feedback pathways in the visual cortex of the macaque monkey (Hof et al. 1995c, 1996). We performed a quantitative analysis of the distribution of neurofilament protein and glutamate receptor subunit (GluRs) proteins GluR5/6/7 (kainate), GluR2/4 (AMPA), and NMDAR1 in corticocortical projection neurons in areas V1, V2, V3, V3A, V4, and MT following injections of the retrogradely transported fluorescent tracers within areas V4 and MT, or in areas V1 and V2 (Tables 2, 3). The laminar distribution of feedforward and feedback projecting neurons is rather heterogeneous. In area V1, only Meynert and layer IVB cells project to area MT. Neurons projecting to area V4 are particularly concentrated in layer III within the foveal representation. In area V2, almost all neurons projecting to areas MT or V4 are in layer III, but are present in both layers II–III and V–VI in areas V3 and V3A. The majority of feedback neurons projecting to areas V1 and V2 are located in layers V–VI in areas V4 and MT, whereas they are found in both layers II–III and V–VI in area V3. A quanti-

Table 2. Neurofilament protein and GluR immunoreactivity in neurons projecting to areas MT and V4[a]

Projection/layer	NFP	GluR5/6/7	GluR2/4	NMDAR1
V1 → MT				
IVB	100	100	100	95.1
Meynert	100	100	100	93.0
V2 → MT				
III	82.6	77.2	98.5	98.2
V	–	–	–	–
V3 → MT				
III	74.5	80.0	100	99.1
V	20.3	54.3	96.9	94.1
V1 → V4				
III	24.1	43.7	100	100
V2 → V4				
III	21.2	48.1	97.6	96.9
V	–	–	–	–
V3 → V4				
III	27.2	53.0	97.4	98.0
V	14.8	35.6	100	100

[a] Results were obtained in seven macaque monkeys and represent pooled percentages of double-labeled neurons in each projection. Note the high prevalence of neurofilament protein (NFP) in the projections to area MT. GluR5/6/7 displays a profile comparable to that of neurofilament protein, whereas NMDA and AMPA receptors are found in almost all of the projecting neurons

Table 3. Neurofilament protein and GluR immunoreactivity in neurons projecting to areas V 1 and V 2[a]

Projection/layer	NFP	GluR5/6/7	GluR2/4	NMDAR1
MT → V1				
III	60.0	80.0	96.6	97.4
V	79.1	86.8	98.8	99.4
V4 → V1				
III	60.1	83.8	98.2	98.6
V	80.1	91.0	99.2	99.8
V3 → V1				
III	64.8	82.4	99.0	99.0
V	72.8	87.0	99.4	99.4
MT → V2				
III	61.8	81.6	98.0	98.0
V	79.6	87.2	99.0	99.0
V4 → V2				
III	63.0	78.5	97.8	97.5
V	79.9	89.0	99.8	99.0
V3 → V2				
III	67.2	80.6	97.8	99.8
V	76.5	85.8	99.2	98.2

[a] Results were obtained in seven macaque monkeys and represent pooled percentages of double-labeled neurons in each projection. Note the overall high prevalence of neurofilament protein (NFP) and GluR5/6/7 in the projections to both areas V 1 and V 2. AMPA and NMDAR1 subunits are present in almost all of the projecting neurons

tative analysis of these projections shows that there are higher numbers of neurons projecting from areas V1, V2, V3, and V3A to area MT that contain neurofilament protein (57–100%) than to area VA (25–36%; Table 2). In contrast, feedback projections from areas MT, V4, and V3 exhibit consistently high proportions of neurofilament protein-immunoreactive neurons (70–80%), regardless of their target areas (in this case areas V1 or V2; Table 3; Hof et al. 1995c, 1996). These results suggest that neurofilament protein identifies particular subpopulations of corticocortically projecting neurons with distinct regional and laminar distribution in the monkey visual system. It is possible that the preferential distribution of neurofilament protein within feedforward connections to area MT and all feedback projections is related to particular physiologic properties of these corticocortical projection neurons. Interestingly, the distribution of the various GluRs in this set of projections indicates that the kainate receptor proteins GluR5/6/7 have a pattern quite similar to that of neurofilament protein, since it is found in approximately 80% of the feedforward connections to area MT and 50% of the projections to area V4, whereas the feedback projections to areas V1 and V2 contain a high proportion of double-labeled neurons (85%; Hof et al. 1995c). In contrast, the GluR2/4 and NMDAR1 subunits are homogeneously and consistently present in nearly all neurons participating in these projections

(95–100 %; Tables 2, 3). This GluR pattern is not ubiquitous, however, since in connections between the prefrontal and temporal cortices, much fewer projections neurons have been reported to contain GluR5/6/7 immunoreactivity (Huntley et al. 1994). Altogether these results suggest that GluR5/6/7 distribution also shows regional and projection-dependent differences, although they appear to be generally more prevalent than neurofilament protein in the visual connections.

The potential interactions among regional specialization, connectivity, and cellular characteristics such as neurochemical profile and morphology have been investigated in a series of experiments involving tract-tracing of motor connections between the cingulate cortex and the putative forelimb region of the primary motor cortex (area M1), cell loading of the retrogradely labeled neurons combined with cell reconstruction, subsequent immunohistochemistry, and confocal laser scanning microscopy to localize neurofilament protein in the identified neurons (Nimchinsky et al. 1996). The cingulate gyrus in monkey and human contains multiple cortical areas that can be distinguished by several neurochemical features, including the distribution of neurofilament protein-enriched pyramidal neurons (Fig. 2C, D). In addition, connectivity and functional properties indicate that there are multiple motor areas in the cortex lining the cingulate sulcus. Two separate groups of neurons projecting to area M1 are conspicuous in the cingulate sulcus, one anterior and one posterior, both of which furnish commissural as well as ipsilateral connections with area M1. The primary difference between the two populations is their laminar origin, with the anterior projection originating largely from the deep layers and the posterior projection emanating from both superficial and deep layers (Nimchinsky et al. 1996). With regard to cellular morphology, the anterior projection exhibits a high degree of morphologic diversity compared to the posterior projection. Commissural projections from both anterior and posterior regions originate principally from layer VI. Neurofilament protein distribution represents a reliable tool for localizing and differentiating these projections. The two sets of projections contain similar proportions of neurofilament protein-immunoreactive neurons, despite the fact that the density and laminar distribution of the total population of neurofilament protein-enriched neurons differs substantially in the two cingulate fields. In these connections, the projecting neurons have a high degree of morphologic heterogeneity (Fig. 6; Nimchinsky et al. 1996). There is generally a high prevalence of true pyramidal neurons in the posterior projection, whereas the anterior projection displays highly diverse cell types, particularly in the deep layers. In these different groups of projection neurons there is no clear correlation between somatodendritic morphology and neurofilament protein content, indicating that the presence of neurofilament protein does not imply homogeneity in morphology (Fig. 6; Nimchinsky et al. 1996). This suggest that at least in the case of these motor connections, neurochemical phenotype may be a more important unifying feature for corticocortical projections than morphology alone. This issue clearly deserves further attention and should be extended to additional sets of projections.

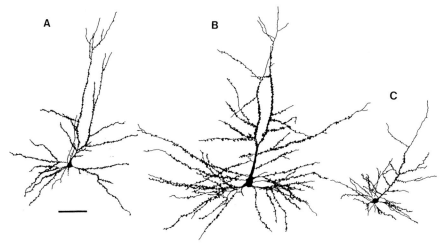

Fig. 6. Reconstructions of neurons loaded with Lucifer Yellow (to reveal their dendritic morphology) participating in the posterior projection from the cingulate cortex to area M1. Cell A is from layer III, cell B from layer V, and cell C from layer VI. The three cells display a typical pyramidal morphology. More unusual shapes were usually seen in layer VI. Only neurons B and C contain neurofilament protein, indicating that there is no correlation between cell morphology and neurofilament protein presence at least in this population of corticocortical neurons. Scale bar = 100 μm. (Adapted from Nimchinsky et al. 1996)

Possible Functions of Neurofilament Protein in Corticocortical Projections

The precise function of neurofilament protein has not yet been determined. However, the facts that this protein has a restricted distribution among certain subsets of corticocortical circuits in the monkey cerebral cortex, and is implicated in the formation of neurodegenerative lesions, suggest that it plays major regulatory roles in a subpopulation of pyramidal neurons particularly vulnerable in dementia. The specific regional and laminar patterns of neurofilament protein-containing neurons also suggest that neurofilament protein confers unique biochemical and structural properties to certain neuronal subpopulations. Thus, in the case of visual connections, neurofilament protein may be crucial for the unique capacity of such subsets of neurons to perform highly precise mapping functions during visual processing. For example, the differential distribution of neurofilament protein-containing neurons among these projections may reflect differential involvement of these neurons in the detection of submodalities, such as color, motion, direction, and speed of movement (Hof and Morrison 1995). Similar conclusions could apply to other primary sensory and motor modalities since in these systems neurofilament protein also exhibits characteristic regional distribution patterns (Campbell and Morrison 1989; Nimchinsky et al. 1996; Hof, unpublished observations).

Several studies have indicated that neurofilament protein is involved in the maintenance and stabilization of the axonal cytoskeleton and that its expression is related to axonal size (Morris and Lasek 1982; Hoffman et al. 1987; Nixon et al. 1994; Pijak et al. 1996). The expression and phosphorylation of the diverse neurofilament protein subunits as well as of other cytoskeletal components during development are neuron-type specific and are involved in determining the adult axonal and somatodendritic cytoskeleton (Riederer et al. 1995, 1996). Our previous results suggest that the presence of high levels of nonphosphorylated neurofilament protein in the somatodendritic compartment of a relatively small number of specialized subsets of neocortical neurons is a major neurochemical feature for maintaining their role in corticocortical networks (Campbell and Morrison 1989; De Lima et al. 1990; Campbell et al. 1991; Hof and Morrison 1995; Hof et al. 1995a, 1996; Nimchinsky et al. 1996). Moreover, the presence of neurofilament protein in the perikaryon has been correlated to cell size, axon size and conduction velocity (Lawson and Waddell 1991). In fact, neurofilament protein is highly represented in retinal magnocellular neurons (Straznicky et al. 1992; Vickers et al. 1995), as well as in the large layer IV B and Meynert cells in area V 1 that furnish a direct projection to area MT (Fries et al. 1985; Hof and Morrison 1995). In this context, it is worth noting that axons of neurons projecting from area V 1 to area MT are larger than axons of other visual projections (Rockland 1989, 1992, 1995). Conversely, the cortical components of the projections to area V display lower levels of neurofilament protein immunoreactivity, indicating differences in neurochemical phenotype in the connections between occipital regions and areas MT and V 4 (Hof et al. 1995a). Definite physiological correlates of functional differences between neurofilament protein-rich and neurofilament protein-poor neurons involved in corticocortical projections are not yet available, although interestingly, conduction velocities have been shown to differ substantially among projection neurons to areas V 4 and MT, such that these neurons projecting to MT fire earlier and have faster conduction (Nowak et al. 1995; Munk et al. 1995). These data warrant further studies of the electrophysiological properties of neurofilament protein-containing neurons.

With respect to cortical processing, the cytoskeletal machinery of the somatodendritic compartment of relatively small neuronal subpopulations that furnish corticocortical projections may represent a phenotypic determinant of the cellular properties underlying the unique computational role of these neurons. For example, it has been reported that the apical dendrites of some motoneurons enriched in neurofilament protein in the cat motor cortex are the target of highly specialized corticocortical inputs involved in learning and consolidation of motor skills (Kaneko et al. 1994). The issue that neurofilament protein-immunoreactive neurons have outstanding functional specificity in corticocortical projections may be related to the facts that there are small numbers of these neurons in the cerebral cortex, estimated to account for approximately 10 to 20 % of all cortical pyramidal neurons (Hof et al. 1990a; Hof and Morrison 1990, 1995. Del Río and DeFelipe 1995), and that combining all cortical projections investigated thus far, approximately 30–40 % of all projection neurons appear to be neurofilament protein-immunoreactive (Hof et al. 1995a). It is therefore possible that the corticocortically projecting neurofilament protein-immunoreactive neurons in layers III and V of neocortical regions in the prefrontal, parietal, and temporal cortices represent a potential cellular substrate of the cortical networks thought to subserve cognitive processing (Goldman-Rakic 1988; Damasio 1989). It has been proposed that a specialized cytoskeleton in particular neuronal subsets may subserve cellular mechanisms involved in synaptic learning rules, attention and awareness, memory binding and consolidation, and ultimately determine the coherent perception and integration of an individual's environment to generate an adapted behavior (Jibu et al. 1994).

Conclusion

Neurofilament protein-immunoreactive neurons are preferentially and dramatically affected by the neurodegenerative process in AD (Morrison et al. 1987; Hof et al. 1990a; Hof and Morrison 1990; Vickers et al. 1992, 1994a), and the cognitive disintegration observed in patients with AD may be due to the loss of long corticocortical projections. Therefore, neurofilament protein-containing neurons may play a central role in the development of the disease (Morrison 1993; Hof and Morrison 1994). Thus, the cytoskeletal destabilization of such neurons may result in a progressive impairment of cognitive and memory functions and ultimately lead to the symptomatology of overt dementia. It is interesting to note, in this context, that transgenic mice expressing the human middle molecular weight subunit of the neurofilament protein exhibit neurofilamentous inclusions resembling NFT and Pick bodies in neocortical pyramidal neurons in an age-dependent manner (Vickers et al. 1994b), and that induction of experimental glaucoma in monkeys results in the degeneration of a subpopulation of large retinal ganglion neurons that selectively contains high neurofilament protein levels (Vickers et al. 1995). Altogether, these results indicate that neurofilament protein-rich neurons are particularly vulnerable in several neurodegenerative conditions, not only in human disease but also in animal models of neurodegeneration, and that the presence of high somatodendritic levels of neurofilament protein may be causally related to the degeneration of key projections neurons.

Also, the apparent association between GluR5/6/7 and neurofilament protein in these pathways may underlie specific physiologic interactions in these neuronal subsets. Studies of the distribution of NMDA and non-NMDA receptor subtypes in the monkey neocortex have shown that specific families of subunits are preferentially associated with distinct pathways (Huntley et al. 1994). The observed differences in GluRs protein distribution suggest that functional interactions may exist at least in certain pathways between neurofilament protein and the high affinity kainate receptor. Also, previous studies of the colocalization of these proteins in the macaque monkey neocortex have demonstrated that the vast majority (>80%) of neurofilament protein-enriched neurons also display

GluR5/6/7 immunoreactivity (Vickers et al. 1993). The similarity in the quantitative distribution of GluR5/6/7 and neurofilament protein in the visual pathways may be quite specific, since analysis of different sets of corticocortical pathways in the prefrontal and cingulate cortices have shown that the distribution of retrogradely labeled GluR5/6/7-immunoreactive neurons is not as strongly correlated to the laminar distribution of neurofilament protein-containing neurons in these regions (Nimchinsky et al. 1993; Huntley et al. 1994). The more ubiquitous distribution of AMPA and NMDA receptor suggests that they are in a position to modulate a large variety of excitatory afferents (Aoki et al. 1994), whereas high affinity kainate receptors could influence, at least in the visual pathways, the function of specialized subsets of corticocortical neurons characterized by high content of neurofilament protein. With respect to neurodegeneration, the fact that excitatory amino acid neurotoxicity induces neuronal alterations comparable to those observed in neurodegenerative conditions, including AD (De Boni and Crapper McLachlan 1985; Choi 1988; Meldrum and Garthwaite 1991; Sautière et al. 1992; Couratier et al. 1996), may be related to a considerable degree to the preferential vulnerability of the neurofilament protein-enriched neurons that furnish certain sets of corticocortical projections.

These morphologic and molecular analyses may be important to develop correlations between distribution of cellular pathologic changes, neurochemical characteristics related to vulnerability and affected cortical circuits. They may also provide the cellular substrate to some of changes observed in functional imaging analyses of the aging human brain. Such integrative approaches could be useful for the development of strategies to prevent or protect against the specific neuronal degenerative events that occur during the progression of dementing illnesses.

Acknowledgments

We thank Drs M. J. Webster, R. Gattass, J. C. Vickers, L. Buée, and A. Delacourte for participation in these studies and helpful discussion, Drs C. Bouras and D. P. Perl for providing postmortem human brain materials, Dr. W. G. Young and his team of programmers for software development, and W. G. M. Janssen, M. M. Adams, C. A. Sailstad, D. M. Blumberg, and R. S. Woolley for expert technical assistance. This work was supported by grants from the NIH (AG06647, AG05138 and the Human Brain Project MHDA52154), the American Health Assistance Foundation, and the Brookdale Foundation.

References

Aoki C, Venkatesan C, Go CG, Mong JA, Dawson TM (1994) Cellular and subcellular localization of NMDA-R1 subunit immunoreactivity in the visual cortex of adult and neonatal rats. J Neurosci 14: 5202–5222

Arnold SE, Hyman BT, Flory J, Damasio AR, Van Hoesen GW (1991) The topographical and neuroanatomical distribution of neurofibrillary tangles and neuritic plaques in the cerebral cortex of patients with Alzheimer's disease. Cereb Cortex 1: 103–116

Benson DF, Davis RJ, Snyder BD (1988) Posterior cortical atrophy. Arch Neurol 45: 789–793

Braak H, Braak E (1991) Neuropathological stageing of Alzheimer-related changes. Acta Neuropathol 82: 239–259

Braak H, Braak E, Kalus P (1989) Alzheimer's disease: areal and laminar pathology in the occipital isocortex. Acta Neuropathol 77: 494–506

Campbell MJ, Morrison JH (1989) A monoclonal antibody to neurofilament protein (SMI-32) labels a subpopulation of pyramidal neurons in the human and monkey neocortex. J Comp Neurol 282: 191–205

Campbell MJ, Hof PR, Morrison JH (1991) A subpopulation of primate corticocortical neurons is distinguished by somatodendritic distribution of neurofilament protein. Brain Res 539: 133–136

Choi DW (1988) Glutamate neurotoxicity and diseases of the nervous system. Neuron 1: 623–634

Clarke S, Miklossy J (1990) Occipital cortex in man: organization of callosal connections, related myelo- and cytoarchitecture, and putative boundaries of functional visual areas. J Comp Neurol 298: 188–214

Couratier P, Lesort M, Sindou P, Esclaire F, Yardin C, Hugon J (1996) Modifications of neuronal phosphorylated τ immunoreactivity induced by NMDA toxicity. Mol Chem Neuropathol 27: 259–275

Damasio AR (1989) The brain binds entities and events by multiregional activation from convergence zones. Neural Comput 1: 123–132

De Boni U, Crapper McLachlan DR (1985) Controlled induction of paired helical filaments of the Alzheimer type in cultured human neurons by glutamate and aspartate. J Neurol Sci 68: 105–118

De Lima AD, Voigt T, Morrison JH (1990) Morphology of the cells within the inferior temporal gyrus that project to the prefrontal cortex in the macaque monkey. J Comp Neurol 296: 159–272

Del Río MR, DeFelipe J (1994) A study of SMI32-stained pyramidal cells, parvalbumin-immunoreactive chandelier cells and presumptive thalamocortical axons in the human temporal necortex. J Comp Neurol 342: 389–408

DeYoe EA, Van Essen DC (1988) Concurrent processing streams in monkey visual cortex. Trends Neurosci 11: 219–225

DeYoe EA, Carman GJ, Bandettini P, Glickman S, Wieser J, Cox R, Miller D, Neitz J (1996) Mapping striate and extrastriate visual areas in human cerebral cortex. Proc Natl Acad Sci USA 93: 2382–2386

Duyckaerts C, Hauw J-J, Bastenaire F, Piette F, Poulain C, Rainsard V, Javoy-Agid F, Berthaux P (1986) Laminar distribution of neocortical senile plaques in senile dementia of the Alzheimer type. Acta Neuropathol 70: 249–256

Felleman DJ, Van Essen DC (1991) Distributed hierarchical processing in the primate cerebral cortex. Cereb Cortex 1: 1–47

Fries W, Kreizer K, Kuypers HGJM (1985) Large layer VI cells in macaque striate cortex (Meynert cells) project to both superior colliculus and prestriate visual cortex area V5. Exp Brain Res 58: 613–636

Goldman Rakic PS (1988) Topography of cognition: parallel distributed networks in primate association cortex. Annu Rev Neurosci 11: 137–156

Grady CL, Maisog JM, Horwitz B, Ungerleider LG, Mentis MJ, Salerno JA, Pietrini P, Wagner E, Haxby JV (1994) Age-related changes in cortical blood flow activation during visual processing of faces and location. J Neurosci 14: 1450–1462

Graff-Radford NR, Bolling JP, Earnest IV F, Shuster EA, Caselli RJ, Brazis PW (1993) Simultanagnosia as the initial sign of degenerative dementia. Mayo Clin Proc 68: 955–964

Haxby JV, Grady CL, Horwitz B, Ungerleider LG, Mishkin M, Carson RE, Herscovitch P, Schapiro MB, Rapoport SI (1991) Dissociation of object and spatial visual processing pathways in human extrastriate cortex. Proc Natl Acad Sci USA 88: 1621–1625

Hof PR, Morrison JH (1990) Quantitative analysis of a vulnerable subset of pyramidal neurons in Alzheimer's disease: II. Primary and secondary visual cortex. J Comp Neurol 301: 55–64

Hof PR, Morrison JH (1991) Neocortical neuronal subpopulations labeled by a monoclonal antibody to calbindin exhibit differential vulnerability in Alzheimer's disease. Exp Neurol 111: 293–301

Hof PR, Morrison JH (1994) The cellular basis of cortical disconnection in Alzheimer disease and related dementing conditions. In: Terry RD, Katzman R, Bick KL (eds) Alzheimer disease. Raven Press, New York, pp 197–229

Hof PR, Morrison JH (1995) Neurofilament protein defines regional patterns of cortical organization in the macaque monkey visual system: a quantitative immunohistochemical analysis. J Comp Neurol 352: 161–186

Hof PR, Bouras C, Constantinidis J, Morrison JH (1989) Balint's syndrome in Alzheimer's disease: specific disruption of the occipito-parietal visual pathway. Brain Res 493: 368–375

Hof PR, Cox K, Morrison JH (1990a) Quantitative analysis of a vulnerable subset of pyramidal neurons in Alzheimer's disease: I. Superior frontal and inferior temporal cortex. J Comp Neurol 301: 44–54

Hof PR, Bouras C, Constantinidis J, Morrison JH (1990b) Selective disconnection of specific visual association pathways in cases of Alzheimer's disease presenting with Balint's syndrome. J Neuropathol Exp Neurol 49: 168–184

Hof PR, Perl DP, Loerzel AJ, Morrison JH (1991a) Neurofibrillary tangle distribution in the cerebral cortex of parkinsonism-dementia cases from Guam: differences with Alzheimer's disease. Brain Res 564: 306–313

Hof PR, Cox K, Young WG, Celio MR, Rogers J, Morrison H (1991b) Parvalbumin-immunoreactive neurons in the neocortex are resistant to degeneration in Alzheimer's disease. J Neuropathol Exp Neurol 50: 451–462

Hof PR, Archin N, Osmand AP, Dougherty GH, Wells C, Bouras C, Morrison JH (1993a) Posterior cortical atrophy in Alzheimer's disease: analysis of a new case and re-evaluation of an historical report. Acta Neuropathol 86: 215–223

Hof PR, Nimchinsky EA, Celio MR, Bouras C, Morrison JH (1993b) Calretinin-immunoreative neocortical interneurons are unaffected in Alzheimer's disease. Neurosci Lett 152: 145–149

Hof PR, Bouras C, Perl DP, Morrison JH (1994) Quantitative neuropathologic analysis of Pick's disease cases: cortical distribution of Pick bodies and coexistence with Alzheimer's disease. Acta Neuropathol 87: 15–124

Hof PR, Nimchinsky EA, Morrison JH (1995a) Neurochemical phenotype of corticocortical connections in the macaque monkey: quantitative analysis of a subset of neurofilament protein-immunoreactive projection neurons in frontal, parietal, temporal, and cingulate cortices. J Comp Neurol 362: 109–133

Hof PR, Mufson EJ, Morrison JH (1995b) The human orbitofrontal cortex: cytoarchitecture and quantitative immunohistochemical parcellation. J Comp Neurol 359: 48–68

Hof PR, Ungerleider LG, Webster MJ, Gattass R, Adams MM, Sailstad CA, Janssen WGM, Morrison JH (1995c) Feedforward and feedback corticocortical projections in the monkey visual system display differential neurochemical phenotype. Soc Neurosci Abstr 21: 904

Hof PR, Ungerleider LG, Webster MJ, Gattass R, Adams MM, Sailstad CA, Morrison JH (1996) Neurofilament protein is differentially distributed in subpopulations of corticocortical projection neurons in the macaque monkey visual pathways. J Comp Neurol 376: 112–127

Hoffman PN, Cleveland DW, Griffin JW, Landes PW, Cowan NJ, Price DL (1987) Neurofilament gene expression: a major determinant of axonal caliber. Proc Natl Acad Sci USA 84: 3472–3476

Huntley GW, Vickers JC, Morrison JH (1994) Cellular and synaptic localization of NMDA and non-NMDA receptor subunits in neocortex: organizational features related to cortical circuitry, function and disease. Trends Neurosci 17: 36–543

Hyman BT, Damasio AR, Van Hoesen GW, Barnes CL (1984) Alzheimer's disease: cell specific pathology isolates the hippocampal formation. Science 225: 1168–1170

Hyman BT, Van Hoesen GW, Kromer LJ, Damasio AR (1986) Perforant pathway changes and the memory impairment of Alzheimer's disease. Ann Neurol 20: 472–481

Hyman BT, Van Hoesen GW, Damasio AR (1990) Memory-related neural systems in Alzheimer's disease: an anatomic study. Neurology 40: 1721–1730

Jibu M, Hagan S, Hameroff SR, Pribram KH, Yasue K (1994) Quantum optical coherence in cytoskeletal microtubules: implications for brain function. BioSystems 32: 195–209

Jones EG, Wise SP (1977) Size, laminar and columnar distribution of efferent cells in the sensory-motor cortex of monkeys. J Comp Neurol 175: 391–438

Kaneko T, Caria MA, Asanuma H (1994) Information processing within the motor cortex. II. Intracortical connections between neurons receiving somatosensory input and motor output neurons of the cortex. J Comp Neurol 345: 161–171

Lawson SN, Waddell JP (1991) Soma neurofilament immunoreactivity is related to cell size and fibre conduction velocity in rat primary sensory neurons. J Physiol 435: 41–63

Levine DN, Lee JM, Fisher CM (1993) The visual variant of Alzheimer's disease: a clinicopathologic case study. Neurology 43: 305–313

Lewis DA, Campbell MJ, Terry RD, Morrison JH (1987) Laminar and regional distribution of neurofibrillary tangles and neuritic plaques in Alzheimer's disease: a quantitative study of visual and auditory cortices. J Neurosci 7: 1799–1808

Livingstone M, Hubel D (1988) Segregation of form, color, movement, and depth: anatomy, physiology, and perception. Science 40: 740–749

Meldrum B, Garthwaite J (1991) Excitatory amino acid neurotoxicity and neurodgenerative diseases. Trends Pharmacol Sci 11: 54–61

Mishkin M, Ungerleider LG, Macko KA (1983) Object vision and spatial vision: two cortical pathways. Trends Neurosci 6: 414–417

Morris JR, Lasek RJ (1982) Stable polymers of the axonal cytoskeleton: the axoplasmic ghost. J Cell Biol 92: 192–198

Morrison JH (1993) Differential vulnerability, connectivity, and cell typology. Neurobiol Aging 14: 51–54

Morrison JH, Lewis DA, Campbell MJ, Huntley GW, Benson DL, Bouras C (1987) A monoclonal antibody to non-phosphorylated neurofilament protein marks the vulnerable cortical in Alzheimer's disease. Brain Res 416: 331–336

Munk MHJ, Nowak LG, Chounlamountri N, Bullier J (1995) Visual latencies in cytochrome oxidase bands of macaque area V2. Proc Natl Acad Sci USA 92: 988–992

Nimchinsky EA, Hof PR, Brose N, Rogers SW, Moran T, Gasic GP, Heinemann S, Morrison JH (1993) Glutamate receptor subunit and neurofilament protein immunoreactivities differentiate subsets of corticocortically projecting neurons in the monkey cingulate cortex. Soc Neurosci Abstr 19: 473

Nimchinsky EA, Vogt BA, Morrison JH, Hof PR (1995) Spindle neurons of the human anterior cingulate cortex. J Comp Neurol 355: 27–37

Nimchinsky EA, Hof PR, Young WG, Morrison JH (1996) Neurochemical, morphologic and laminar characterization of cortical projection neurons in the cingulate motor areas of the macaque monkey. J Comp Neurol 374: 136–160

Nixon RA, Paskevich PA, Sihag RK, Thayer CY (1994) Phosphorylation on carboxyl terminus domains of neurofilament proteins in retinal ganglion cell neurons in vivo: influences on regional neurofilament accumulation, interneuronal spacing, and axonal caliber. J Cell Biol 126: 1031–1046

Nowak LG, Munk MHJ, Girard P, Bullier J (1995) Visual latencies in areas V1 and V2 of the macaque monkey. Vis Neurosci 12: 371–384

Pearson RCA, Esiri MM, Hiorns RW, Wilcock GK, Powell TPS (1985) Anatomical correlates of the distribution of the pathological changes in the neocortex in Alzheimer disease. Proc Natl Acad Sci USA 82: 4531–4534

Pijak DS, Hall GF, Tenicki PJ, Boulos AS, Lurie DI, Selzer ME (1996) Neurofilament spacing, phosphorylation, and axon diameter in regenerating and uninjured lamprey axons. J Comp Neurol 368: 569–581

Riederer BM, Draberova E, Viklicky V, Draber P (1995) Changes of MAP2 phosphorylation during brain development. J Histochem Cytochem 43: 1269–1284

Riederer BM, Porchet R, Marugg RA (1996) Differential expression and modification of neurofilament triplet proteins during cat cerebellar development. J Comp Neurol 364: 704–717

Rockland KS (1989) Bistratified distribution of terminal arbors of individual axons projecting from area V1 to middle temporal area (MT) in the macaque monkey. Vis Neurosci 3: 155–170

Rockland KS (1992) Configuration, in serial reconstruction, of individual axons projecting from area V2 to V4 in the macaque monkey. Cereb Cortex 2: 353–374

Rockland KS (1995) Morphology of individual axons projecting from area V 2 to MT in the macaque. J Comp Neurol 355: 15–26

Rogers J, Morrison JH (1985) Quantitative morphology and regional and laminar distributions of senile plaques in Alzheimer's disease. J Neurosci 5: 2801–2808

Sautière PE, Sindou P, Couratier P, Hugin J, Wattez A, Delacourte A (1992) Tau antigenic changes induced by glutamate in rat primary culture model: a biochemical approach. Neurosci Lett 140: 206–210

Senut MC, Roudier M, Davous P, Fallet-Bianco C, Lamour Y (1991) Senile dementia of the Alzheimer type: is there a correlation between entorhinal cortex and dentate gyrus? Acta Neuropathol 82: 306–315

Straznicky C, Vickers JC, Gábriel R, Costa M (1992) A neurofilament protein antibody selectively labels a large ganglion cell type in the human retina. Brain Res 582: 123–128

Trojanowski JQ, Schmidt ML, Shin RW, Bramblett GT, Rao D, Lee VMY (1993) Altered *tau* and neurofilament proteins in neurodegenerative diseases: diagnostic implications for Alzheimer's disease and Lewy body dementias. Brain Pathol 3: 45–54

Ungerleider LG (1995) Functional brain imaging studies of cortical mechanisms for memory. Science 270: 769–775

Van Essen DC, Maunsell JHR (1983) Hierachical organization and functional streams in the visual cortex. Trends Neurosci 6: 370–375

Vickers JC, Delacourte A, Morrison JH (1992) Progressive tranformation of the cytoskeleton associated with normal aging and Alzheimer's disease. Brain Res 594: 273–278

Vickers JC, Huntley GW, Edwards AM, Moran T, Rogers SW, Heinemann SF, Morrison JH (1993) Quantitative localization of AMPA/kainate and kainate glutamate receptor subunit immunoreactivity in neurochemically identified subpopulations of neurons in the prefrontal cortex of the macaque monkey. J Neurosci 13: 2981–2992

Vickers JC, Riederer BM, Marugg RA, Buée-Scherrer V, Buée L, Delacourte A, Morrison JH (1994a) Alterations in neurofilament protein immunoreactivity in human hippocampal neurons related to normal aging and Alzheimer's disease. Neuroscience 62: 1–13

Vickers JC, Morrison JH, Friedrich Jr VL, Elder GA, Perl DP, Katz RN, Lazzarini RA (1994b) Ageassociated and cell-type-specific neurofibrillary pathology in transgenic mice expressing the human midsized neurofilament subunit. J Neurosci 14: 5603–5612

Vickers JC, Schumer RA, Podos SM, Wang RF, Riederer BM, Morrison JH (1995) Differential vulnerability of neurochemically identified subpopulations of retinal neurons in a monkey model of glaucoma. Brain Res 680: 23–35

Watson JDG, Myers R, Frackowiak RSJ, Hajnal JV, Woods RP, Mazziotta JC, Shipp S, Zeki S (1993) Area V 5 of the human brain: evidence from a combined study using positron emission tomography and magnetic resonance imaging. Cereb Cortex 3: 79–94

Reduced Neuronal Activity is One of the Majc Hallmarks of Alzheimer's Disease

D. F. Swaab[*], *P. J. Lucassen, J. A. P. van de Nes, R. Ravid and A. Sc*

Introduction

Alzheimer's disease (AD) is histopathologically characterized by the presence of neuritic plaques (NPs) and cytoskeletal changes (Fig. 1) that are visible as pretangles stained by Alz-50, neurofibrillary tangles (NFT) in the cell bodies of affected neurons, and neuropil threads (Braak et al. 1986) or dystrophic neurites (Kowall and Kosik 1987). Dystrophic neurites are defined as short, thickened, curly, coiled or sometimes hooked fibres observable as the neuritic component of NPs or present in the neuropil outside these structures. Neuropil threads is an alternative term for dystrophic neurites that are not the neuritic component of NPs. To a lesser degree, NPs and cytoskeletal changes can also be observed in aged, nondemented control subjects.

This review will provide evidence in favour of the thesis that 1) the neuropathological Alzheimer changes cannot all be explained by a cascade starting with amyloid deposits as suggested, e.g., by Selkoe (1994), 2) the neuropathological hallmarks of AD are basically independent phenomena and 3) neuronal inactivity is one of the major characteristics of AD and may underlie the clinical symptoms of dementia.

Amyloid: Just one of the Many Factors Involved

The amyloid cascade hypothesis, as advocated by Selkoe (1994), constitutes a major working hypothesis in current AD research. Amorphous plaque deposition is thought to be the primary event. The congophilic components of NPs are assumed to arise from the congo-negative amorphous plaques by aggregation of β/A4-protein fibrils. Due to the neurotoxicity of β/A4, NP formation, the occurrence of neuropil threads and dystrophic neurites would be induced followed by NFT formation and ultimately cell death. The toxic potency of amorphous plaque deposits is attributed to the β/A4-protein (Masters et al. 1985) that consists of 40–43 amino acids that are cleaved from the much larger amyloid precursor protein (APP; Kang et al. 1987). Grafting genetically transformed cells that over-

[*] Netherlands Institute for Brain Research, Meibergdreef 33, 1105 AZ Amsterdam, The Netherlands

B. T. Hyman / C. Duyckaerts / Y. Christen (Eds.)
Connections, Cognition, and Alzheimer's Disease
© Springer-Verlag Berlin Heidelberg 1997

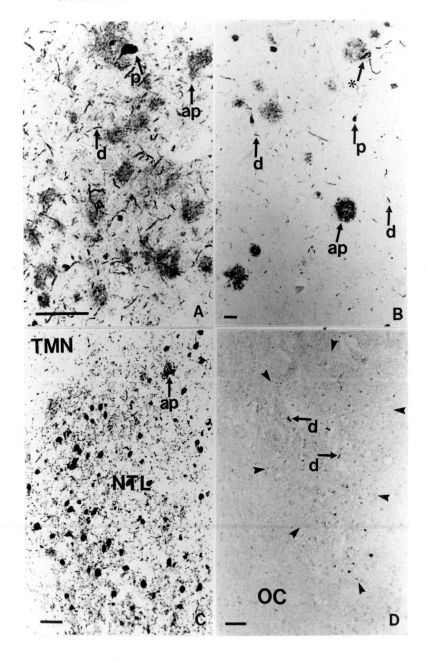

◀ **Fig. 1.** Localization of cytoskeletal changes and amorphous plaques. A. Alzheimer's disease (AD) patient, 90 years of age. A4-reactive plaque deposits (ap) and Alz-50-stained dystrophic neurites (d) and perikarya (p) are found close together in the seriously affected AD frontal cortex. The A4-antiserum used for this staining was SP28. B. AD patient, 40 years of age. The occurrence of amorphous plaques (ap) and dystrophic neurites (d) is more discrete in the AD hypothalamus than in the AD cerebral cortex. Although occasionally Alz-50 positive dystrophic neurites can be found apposed to amorphous plaques (*), cytoskeletal changes are mostly not locally accompanied by plaque deposits. C. AD patient, 70 years of age. The nucleus tuberalis lateralis (NTL) filled with Alz-50-stained cytoskeletal changes can easily be distinguished from the tuberomamillary nucleus (TMN), which shows a considerably lower density of AD-related altered cytoskeletal structures. Note that the vast majority of cytoskeletal changes are not accompanied by A4-reactive plaque deposits (ap). The A4 antiserum used was SP28. D. AD patient, 40 years of age. Alz-50-stained dystrophic neurites (d) in the suprachiasmatic nucleus (OC, optic chiasm) were never observed to coexist with SP28-reactive plaque deposits (arrowheads indicate the borders of the suprachiasmatic nucleus). All bars represent 100 μm, except that in Fig. 1B, which represents 50 μm (from Van de Nes et al. 1994, with permission).

---

express β/A4 amyloid into the suprachiasmatic nucleus of adult rats indeed caused a disruption of circadian rhythms, a finding that favour the idea that this compound is neurotoxic (Tate et al. 1992). Moreover, AD is linked to trisomy 21 (Down syndrome) and in some cases to a point mutation in the APP gene on chromosome 21. Although there are certainly arguments in favour of Selkoe's cascade hypothesis, there are also data from a large number of papers that do not fit into this hypothesis (for review see Van de Nes et al. 1994). The vast majority of AD cases (99.5%) are not linked to APP gene mutations or an extra copy of chromosome 21 (Rossor 1993). The occurrence of familial AD cases linked to chromosomes other than chromosome 21, e.g., chromosome 14 (Mullan et al. 1992; Schellenberg et al. 1992; St George-Hyslop et al. 1992; Van Broeckhoven et al. 1992), or 1 (Levy-Lahad et al. 1995) as well as the cases linked to chromosome 19 (Pericak-Vance et al. 1991; Strittmatter et al. 1993 a, b) cannot be explained in a direct way by the amyloid cascade hypothesis.

Cytoskeletal Changes without Senile Plaques

The histopathological changes in the punch-drunk dementia syndrome of boxers (dementia pugilistica) also do not fully agree with the amyloid cascade hypothesis. The hippocampus and cortex of AD-demented boxers show many amorphous plaques, NFTs and neuropil threads, but very few if any NPs (Roberts et al. 1990). In several cases of Alzheimer patients, Bouras et al. (1994) found NFTs without amyloid deposition. These observations indicate that cytoskeletal changes may occur independent of aggregated β/A4 fibril amyloid cores.

No Transformation of Amorphous to Neuritic Plaques (NPs)

It is presumed by many authors that there is a transformation from amorphous plaques to NPs, but the relationship between these AD changes is equivocal both in histopathological and biochemical terms. Amorphous plaques do not seem to be neurotoxic and induce NFT. Cytoskeletal changes and amorphous plaques are not related in their localization (Fig. 1). In the hippocampus and neocortex of aged, nondemented control subjects, large amounts of amorphous plaques may be present, whereas NFTs are practically absent (Duyckaerts et al. 1988; Barcikowska et al. 1989; Delaere et al. 1990) or the presence of neuropil threads is low (Dickson et al. 1988; McKee et al. 1991). Amorphous plaques are present in the cerebellum of AD and Down syndrome patients without the presence of neurofibrillary changes in the brain area (Braak et al. 1989a; Joachim et al. 1989). In addition, the widespread presence of amorphous plaques is not necessarily associated with AD dementia. The percentage of aged, nondemented subjects over 75 years of age with β/A4 deposits in the cortex may reach 80% (Davies et al. 1988), whereas for the parahippocampal and superior temporal gyri of nondemented centenarians this proportion may even approximate 100% (Delaere et al. 1993). In addition, large amounts of amorphous plaques may also be found with hardly any neurofibrillary changes in the parvocellular layer of the presubiculum (Kalus et al. 1989; Akiyama et al. 1990). Also, Bouras et al. (1994) found that the extent of amyloid deposition in the hippocampal area was not correlated with the diagnosis of Alzheimer's disease.

Amorphous plaque deposition in the brain is rather diffuse, whereas SPs are concentrated in particular areas. Amorphous plaques are found in all six neocortical layers (Braak et al. 1989b; Arnold et al. 1991), whereas classical and NPs are mainly found in layers II–III (Pearson et al. 1985; Rogers and Morrison 1985; Duyckaerts et al. 1986).

Furthermore, amorphous plaques also occur in 66% of the elderly (i.e., over 65 years of age) affected by neurodegenerative disorders other than AD, e.g., progressive supranuclear palsy, Parkinson's disease, Huntington's chorea and frontal lobe dementia. However, amorphous plaques are widely associated with NPs and neuropil threads in AD only (Mann and Jones 1990).

The transformation from amorphous plaques into NPs can also be questioned on the basis of the finding that amorphous plaques stain with antibodies raised against the β/A4-protein and the APP extracellular epitope, but not with antisera raised against the APP carboxy-terminal region, whereas classical NPs stain with all three antisera (Tagliavini et al. 1991). These observations indicate that there may also be differences in the chemical composition of classical NPs and amorphous plaques in addition to a difference in the aggregated versus the nonaggregated state of the β/A4-protein fibrils.

The appearance of plaques that might represent intermediate forms between amyloid plaques and NPs has never been reported.

Neuritic Plaques and Neurofibrillary Tangles may Occur Independently

The NPs with a congophilic amyloid core have obtained particular attention, since amyloid has a central place in the cascade hypothesis. However, it is not the NPs but the cytoskeletal changes that parallel the duration and severity of AD (Arriagada et al. 1992; Braak and Braak 1991; McKee et al. 1991; Mukaetova-Ladinska et al. 1993). NFTs are often proposed to have their starting point in the neurotoxicity of NPs. This idea is supported by the observation that many classical NPs are present in the "terminal zone of the perforant pathway," which corresponds to the outer two-thirds of the molecular layer of the dentate gyrus and to the distal dendrites of the subicular CA1 zone (Hyman et al. 1986). The neurons of the perforant pathway projections that originate from the entorhinal cortex in fact contain many NFTs in AD (Hyman et al. 1986). However, the idea that the formation of NFTs in entorhinal cortex neurons and of NPs in the hippocampal CA1 region is necessarily linked is not supported by statistical analysis. The correlation between the density of cellular NFTs in the entorhinal cortex neurons and classical NPs in the dentate gyrus and CA1 region as projection area appears to be weak (Armstrong et al. 1992). In addition, clusters of NPs and tangles are spatially not related (Armstrong et al. 1993). Moreover, 30 % of the demented senile "Alzheimer patients" with plaques in the neocortex appeared to lack tangles in this brain area (Terry et al. 1987). This finding suggests that the occurrence of classical SPs on the one hand and NFT formation on the other is rather independent.

There is also no precise relationship between the clustering of NPs in the neocortex on the one hand and the localization of the terminals of the magnocellular cholinergic neurons of the basal forebrain or noradrenergic cells in the locus coeruleus that contain the NFT. In the neocortex, NPs predominantly occur in layers II–III (Pearson et al. 1985; Rogers and Morrison 1985; Duyckaerts et al. 1986), whereas data on cholinesterase histochemistry show the most dense staining in layers I and VI (Mesulam et al. 1983). These are precisely the neocortical laminae that have the fewest NPs. This means that the layers to which the magnocellular cells of nucleus basalis of Meynert (NBM) project do not coincide with the layers with the densest NP concentration. In addition, dopamine-β-hydroxylase histochemistry shows that the locus coeruleus neurons also mainly send their axons to layers I and VI (Morrison et al. 1982). Thus, the neocortical layers to which locus coeruleus neurons send their axons do not correspond with the neocortical layers that show a concentration of NPs, i.e., layers II and III. Therefore, NFT formation in the cholinergic neurons of the NBM and the noradrenergic cells of the locus coeruleus appears to be independent of NP deposition in the neocortex.

The possibility of independent formation of NPs and NFT is further supported by the reported cases of AD that show a lack of NFTs (Terry et al. 1987).

Similar discrepancies are found in the striatum. AD-affected type IV and V interneurons in the caudate nucleus and putamen develop NFTs and dystrophic neurites, but hardly any classical NPs are found (Braak and Braak 1990). This indicates that NFT formation in the striatum is a local pathology not induced by classical NP deposition in this structure.

Another histopathologic argument that suggests that cytoskeletal changes are not necessarily locally induced by aggregated β/A4-protein fibrils is provided by our observations in the hypothalamus. Firstly, congophilic plaques in the hypothalamus are rare (Standaert et al. 1991). However, various hypothalamic nuclei appear to be affected by Alz-50-stained "pretangle" cytoskeletal changes (Swaab et al. 1992; Van de Nes et al. 1993). In addition, the hypothalamic nucleus tuberalis lateralis (NTL) undergoes hardly any amorphous or NP formation in AD (Kremer et al. 1991; Van de Nes et al. 1996), although the nucleus stains intensively with Alz-50-stained cytoskeletal changes (Kremer et al. 1991; Swaab et al. 1992; Van de Nes et al. 1993; see Fig. 1). The hypothalamic suprachiasmatic nucleus may show some Alz-50-stained cell bodies and altered fibres in AD, but amorphous or classical plaques have never been observed in this nucleus (Fig. 1).

These observations suggest that induction of pretangles, neuropil threads and NFTs by neurotoxic NPs does not seem to play a crucial role in AD development, as proposed in the amyloid cascade hypothesis. Rather, NP formation has to be seen as an independent process. This would also explain why no spatial relationship is seen between the distribution of neuropil threads and NPs. In addition, it would clarify why large amounts of NPs can be found in the neocortex of nondemented patients, whereas neuropil threads (Dickson et al. 1988; McKee et al. 1991) or NFTs are virtually absent (Katzman et al. 1987; Crystal et al. 1988; Arriagada et al. 1992).

Cytoskeletal Changes and NPs do not Induce Cell Death

The last step of the amyloid cascade hypothesis is the idea that neurons producing cytoskeletal changes are indicative of impending neuronal death and that cell death would be induced by the neurotoxic NPs.

The findings in the AD locus coeruleus are in agreement with this last part of the hypothesis. Large neuromelanin pigment-containing neurons in the locus coeruleus develop NFTs, and neuronal death may be up to 80 % (Bondareff et al. 1982; Chan-Palay and Asan 1989; German et al. 1992; Hoogendijk et al. 1995). The proposed sequence, i.e., that NFT formation is followed by dramatic cell loss, was thought to be common AD. However, this only holds for a few brain areas (e.g., locus coeruleus and CA1 of the hippocampus). Cell loss is indeed considerable in the latter structure in AD (West et al. 1994). This is in contrast to the NBM, in spite of the fact that in early reports, neuronal loss of up to 90 % has also been reported (Mann et al. 1984; Whitehouse et al. 1982), only a small part of which was related to aging (Lowes-Hummel et al. 1989; De Lacalle et al. 1991). However, the reported loss in the large-sized cholinergic neurons in the NBM in AD appears to be related to cell shrinkage, reduced activity, and loss of cholinergic cell markers rather than to cell loss (Rinne et al. 1987; Allen et al. 1988;

Salehi et al. 1994). Regeur et al. (1994), using unbiased sampling and counting methods, showed that global neocortical cell loss does not take place in this brain area of AD patients, providing strong evidence that neuronal shrinkage rather than cell death is a major phenomenon in AD.

Data on Alz-50-stained cytoskeletal changes and neuronal cell counts of the sexually dimorphic nucleus (SDN) and NTL in control and AD hypothalami of different ages do not support the idea that neurons staining for cytoskeletal changes are indicative of impending cell death. Alz-50-stained cell bodies and dystrophic neurites in the SDN have only been observed in AD and not in controls (Swaab et al. 1992; Van de Nes et al. 1993), whereas the pattern of loss of SDN neurons in AD is similar to that during normal aging (Swaab and Hofman 1988). In addition, the Alz-50 staining of the NTL of AD patients is so abundant that it can even be seen with the unaided eye, although neuron number in AD is not different from that in control subjects (Kremer et al. 1991).

The CA1 area of the hippocampus is one of the few areas that shows a very clear neuronal loss in AD. West et al. (1994) have shown that the number of neurons in the CA1 area of the hippocampus is much more reduced in AD than in normal aging. Since this area shows not only a massive number of NFTs but also NPs in AD, the CA1 was chosen to study the question of whether or not SPs may induce local cell death (Emre et al. 1992; Kowall et al. 1992). Our study (Salehi et al., submitted for publication) shows that there is indeed a slightly lower neuronal density around NPs. In addition, we found a negative relationship between the size of the neuritic plaques and neuronal density around them, indicating that this effect was dose dependent. However, it should be noted that the contribution of this effect on the total cell death in the CA1 area is very limited, i.e., 2,6 %. This study again supports the notion that SPs and cell death are largely two independent phenomena.

No Animal Models for AD

Based on observations on mutations in the APP gene in some familial AD cases and early-onset AD dementia in trisomy 21 Down syndrome patients that is presumed to be related to overproduction of normal APP, various transgenic mouse models overexpressing APP have been developed and tested. However, the results obtained thus far are rather disappointing. A few years ago, a major step forward was claimed by Kawabata et al. (1991), who reported the detection of amyloid deposits and NFTs in the transgenic mouse brain. However, the authors had to retract their paper since they could not reproduce the histopathological AD changes. Another research group reported the presence of small deposits of age-related β/A4-protein-like immunoreactivity in the mouse hippocampus (Wirak et al. 1991). These deposits, however, appeared to be clusters of intracytoplasmic inclusions in astrocytic processes, i.e., a murine corpora amylacea-like structure that can exhibit nonspecific staining patterns with a variety of polyclonals (Jucker et al. 1992). Other transgenic mouse models with an overproduction of normal APP revealed some $\beta/A4$-amyloid-like deposits in the hippocampus and cortex, but significant behavioral changes or additional histopathological changes associated with AD were not found (Quon et al. 1991; Kammesheidt et al. 1992; Andrä et al. 1996; Games et al. 1995; Malherbe et al. 1996). Even the best animal model, described by Games' group (Greenberg et al. 1996) using an APP mutation that gives a 10-fold increase in the level of APP, did not show NFTs (Games et al. 1995). In addition, human neurons that secrete $A\beta$ did not induce Alzheimer pathology following transplantation into rodent brain (Mantione et al. 1995). Thus, so far, no animal model has yet been developed that can support the amyloid cascade hypothesis.

Decreased Neuronal activity is a Major Hallmark of Alzheimer's Disease

It has been proposed that decreased neuronal activity is an essential characteristic of AD and that neuronal activity would protect against the degenerative changes of aging and AD, an hypothesis paraphrased as "use it or lose it" (Swaab 1991). In fact, there are studies in the literature supporting the view that decreased metabolic rate is a major hallmark of AD. It has been reported that the AD brain shows a lower total amount of protein (Suzuki et al. 1965), a clear reduction in total cytoplasmic (Bowen et al. 1977; Mann et al. 1981; Doebler et al. 1988) and messenger RNA (Sajdel-Sulkowska and Marotta 1994; Guillemette et al. 1986; Taylor et al. 1986), reduced glucose metabolism especially in temporal and parietal lobes, as shown by positron emission tomography (Hoyer et al. 1988; Kumar et al. 1993; Meneilly and Hill 1993; Meier-Ruge et al. 1994; Swerdlow et al. 1994), a smaller size of the neuronal Golgi apparatus (Salehi et al. 1994, 1995b, c) and a lower cytochrome oxidase level (Simonian and Hyman 1993, 1994). In this respect it is also interesting to note that isolated microvessels from Alzheimer patients' temporal cortex showed decreased glucose metabolism, suggesting a global defect in brain energy metabolism (Marcus et al. 1989).

Foster et al. (1984) have demonstrated that a substantial decrease in cerebral glucose metabolism may precede significant cognitive impairment. This observation was supported by Reiman et al. (1996), who found that late middle-aged, cognitively normal subjects who are homozygous for the APOE E4 allele, and thus at risk for AD, have already reduced glucose metabolism in the same region of the brain as is later affected in patients with probable AD. Reduction of regional cerebral glucose metabolism in later stages of AD is related to neuropsychological impairment (Haxby et al. 1988; Mielke et al. 1994). It is interesting to note that Parkinson patients with dementia also show a global decrease in glucose metabolism similar to that in AD, i.e., with more severe abnormalities in the tempero-parietal region (Peppard et al. 1992). It is presumed that cortical glucose hypometabolism in AD may reflect reduced synaptic activity (Salmon et al. 1996).

Two major questions concerning the pathogenesis of AD were therefore 1) whether the presence of plaques and tangles in AD was indeed related to decreased neuronal activity in the various brain areas, and, if so, 2) whether these neuropathological AD hallmarks would induce decreased metabolic rate or vice versa, or alternatively whether the neuropathological AD changes and decreased metabolic rate would occur independently. Our recent research supports the latter possibility.

Relationship Between Alzheimer Neuropathology and Decreased Metabolism

Using the size of the Golgi apparatus as a histological parameter of chronic activity changes in neurons, we studied the relationship between the activity of neurons in areas with different types of neuropathological AD changes. The hypothalamus contains several nuclei that are differentially affected in AD. For instance, the supraoptic nucleus (SON) is not affected at all by AD changes and even shows hyperactivation during aging, both in controls and AD patients (Swaab et al. 1992; Lucassen et al. 1994). This is in contrast to the NBM, which shows clear signs of atrophy (Rinne et al. 1987), cytoskeletal alterations (Swaab et al. 1992; Van de Nes et al. 1993) and some NP formation (Rudelli et al. 1984). In addition, we studied the CA1 area of the hippocampus, a brain region that is affected by cell death (West et al. 1994), an abundance of NFTs and a moderate amount of NPs (Mann et al. 1985). In the hypothalamic and hippocampal tissue from controls and AD patients, we investigated the possible presence or absence of a relationship between different stages of neurofibrillary degeneration and neuronal activity and wondered what could be cause or effect of such a relationship. In order to do so, the relationship between Golgi apparatus size and the following stages of AD changes were assessed in case of:

No AD changes

The SON of the hypothalamus appears to be completely spared in AD. No classical AD neuropathology is present and, even using antibody Alz-50 as an indicator of early cytoskeletal alterations (Bancher et al. 1989), no staining of SON neurons is found (Swaab et al. 1992; Van de Nes et al. 1993). Furthermore, no cell loss in the SON is found either in aging or AD (Goudsmit et al. 1990; Van der Woude et al. 1995). In contrast, even signs of hyperactivation with aging were found in the SON (Hoogendijk et al. 1995; Van der Woude et al. 1995). As shown by Lucassen et al. (1994; Fig. 2), there was indeed a significant increase in activity of vasopressinergic neurons of the SON during aging, both in controls and AD patients, supporting the idea that activation of neurons might protect them against AD changes (Swaab 1991).

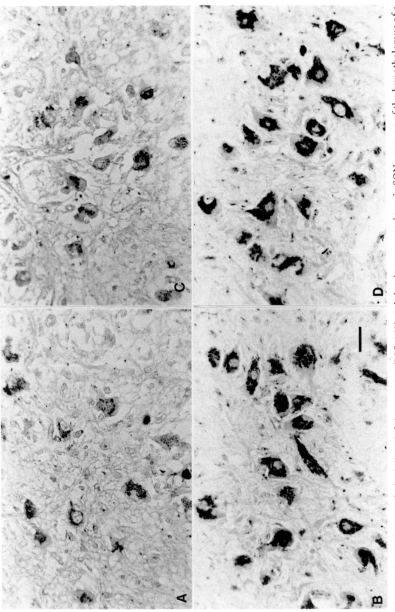

Fig. 2. Photomicrograph depicting Golgi apparatus (MG-160) staining in vasopressinergic SON neurons of the hypothalamus of a 43-years-old control subject (A), an 82-years-old control subject (B), a 49-year-old Alzheimer's patient (C) and an 81-year-old Alzheimer's patient (D). Note the increase in size of the Golgi apparatus with age. Bar represents 28 µm (from Lucassen et al. 1994, with permission).

Early Cytoskeletal Alterations

Tau proteins belong to the microtubule associate proteins. Abnormal phosphory-lation of tau proteins as found in AD patients may lead to failure of tau proteins to bind to microtubules (Lee et al. 1991). This would result in a failure to main-tain axonal transport and neuronal shape. It has been suggested that the occur-rence of early cytoskeletal changes due to abnormal phosphorylation of tau, as found by a variety of antibodies, would precede the appearance of neurofibrillary degeneration, as shown by silver staining (Bancher et al. 1989; Braak et al. 1994).

The hypothalamic nucleus tuberalis lateralis (NTL) shows strong cytoskeletal alterations, as appears from the intense staining of NTL neurons by the anti-bodies Alz-50, tau-1 (against tau) and 3–39 (against ubiquitin; Swaab et al. 1992; Van de Nes et al. 1993; Fig. 1). Interestingly, silver stained NFTs and NPs are rare in the NTL of AD brains (cf., Kremer 1992). The NTL thus represents a brain area that shows an early stage of AD changes and does not progress towards classical silver staining of neuropathological AD hallmarks (Swaab et al. 1992), which made it a very suitable structure for a study of the relationship between the pres-ence of pretangles and changes in neuronal activity. As shown in Salehi et al. (1995a), there was no reduction in the activity of this area in AD (Fig. 3). Fur-thermore, comparison of the intensity of Alz-50 staining with Golgi apparatus (GA) size did not show a clear relationship between these two parameters. This study showed that strong cytoskeletal alterations in the NTL are not accom-panied by decreased neuronal activity and that pretangle AD changes and reduced metabolic activity are not necessarily related.

Late Cytoskeletal Alterations (NFTs)

The appearance of early cytoskeletal alterations is presumed to be followed by the formation of NFTs that are detectable by silver staining (Bancher et al. 1989). The finding (see above) that there was no relationship between the appearance of early cytoskeletal alterations and protein synthetic ability in the NTL raised the question whether AD changes and decreased metabolism would be related in areas with late stages of cytoskeletal alterations, i.e., NFTs and SPs such as the NBM, tuberomamillary nucleus (TM) and CA1 area of the hippocampus.

The NBM is an area of the basal forebrain that is severely affected in AD. This nucleus shows not only early cytoskeletal alterations, as indicated by Alz-50 staining, but also NFTs and β-amyloid accumulation and NPs in AD (Rudelli et al. 1984; Swaab et al. 1992; Van de Nes et al. 1993). Although it was suggested initially that this area shows dramatic cell death in AD (Whitehouse et al. 1982; 1983), it turned out that degeneration in the NBM is characterized by cell atrophy rather than by cell death (Pearson et al. 1983; Rinne et al. 1987). A significantly decreased size of the GA was found in NBM neurons in AD, suggesting that pro-tein synthetic activity of NBM neurons is strongly reduced in this brain area (Salehi et al. 1994; Fig. 4).

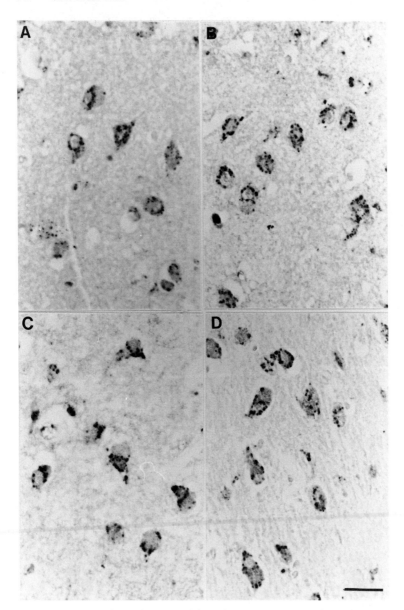

Fig. 3. Immunocytochemical staining of the Golgi apparatus (GA) in the nucleus tuberalis lateralis of a young control (A) and old control (B) and young AD patient (C) and old AD patient (D). Note the similarity between the GA of cells of the four groups. Scale bar 30 μm (from Salehi et al. 1995a, with permission).

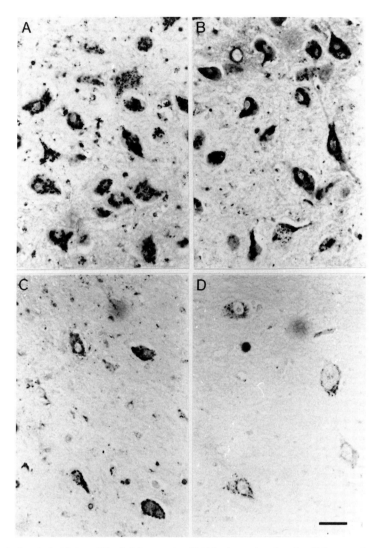

Fig. 4. Immunocytochemical staining of the Golgi apparatus (GA) in the nucleus basalis of Meynert of a young (A) and old (B) control and young and old AD patients (C, D). Note the clear reduction in size of the GA in AD patients when compared to the controls (from Salehi et al. 1994, with permission).

As shown by Salehi et al. (1995c), metabolic activity of TM neurons, an area of the hypothalamus that is clearly affected by NFTs (Nakamura et al. 1993), is also significantly reduced in AD, which supported again the existence of a relationship between AD pathology and decreased neuronal activity. The same holds for the CA1 area of the hippocampus that is strongly affected by AD changes (West et al. 1994) and where, as shown by Salehi et al. (1995b), neuronal activity was strongly decreased in AD patients.

In conclusion, the pretangle stage of AD changes is not necessarily related to changes in metabolic rate as indicated by our studies in the NTL, However, both 1) the clear reduction in Golgi apparatus size in the AD affected NBM, TM and CA1 neurons and 2) the significant increase in activity of the vasopressinergic neurons during aging and AD in the SON, an area where no cytoskeletal abnormalities are found, support the idea that decreased neuronal activity and the occurrence of the classical AD changes go together. Yet, these observations could not reveal the causality of such a relationship.

Tangles and Neuritic Plaques (NPs) do not Cause Decreased Metabolic Rate

The next step was to study the causality of the relationship between the presence of NPs and NFT in a brain area with decreased metabolic activity. For this purpose we compared the metabolic rate of CA1 neurons that did contain NFTs with those that did not. There appeared to be no difference in the size of the Golgi apparatus between these two groups of neurons. The presence of NFT does not seem to decrease the metabolic rate of a neuron (Salehi et al. 1995b). So, although NFT and decreased metabolic activity are present in the same brain area, i.e., CA1, they do not seem to be causally related. This finding agrees with the observation of Gertz et al. (1989), who showed that the presence of intraneuronal NFT in the CA1 area of the hippocampus is not related to other parameters of metabolic activity, i.e., nucleolar or cell size.

NPs are considered by some to be later stages of amorphous plaques (Rozemuller et al. 1989). Because of extensive damage to the neuropil in the area of NPs, they are also called "malignant" plaques (Wisniewski and Wegiel 1995). Although it is still a matter of controversy, many investigators believe that the β-amyloid content of the core of the plaques is neurotoxic and induces neural degeneration. On the other hand, unlike NFTs, there is no clear relationship between the number of NPs and the severity of dementia (see before), which makes a neurotoxic effect of plaques as a major pathogenetic mechanism in AD questionable. If a plaque contained neurotoxic compounds, one would expect that the closer a neuron is situated to the plaque, the lower its metabolic rate would be. Our measurements do not support the importance of such a mechanism. There appeared to be no relationship between either the density of NPs or the distance of each NP to the activity of neighbouring neurons (Salehi et al., submitted for publication). This finding does not support the possibility that neurotoxicity of plaques caused decreased neuronal metabolism but rather that metabolism and NPs are two basically independent phenomena.

Restoration of Neuronal Activity

The present review showed that there is a clear reduction in neuronal metabolic activity in various brain areas in AD patients. On the basis of this important hallmark of AD (Swaab 1991), one may assume that restoration of the activity of neurons would lead to diminishment of cognitive impairment. Although it is not yet clear whether decreased metabolic activity is a primary process in the course of AD, it has been shown that, in principle, reactivation of neurons might be beneficial for AD patients. Global stimulation by exercise, transcutaneous electric nerve stimulation (TENS), tactile nerve stimulation and a combination of the latter two significantly improved memory and affective behaviour of patients with probable AD (Powell 1974; Scherder et al. 1995a–c). Fragmentation of sleep-wake patterns occurs in senescence and even more clearly in AD (Witting et al. 1990), which is believed to be due to degenerative changes in the suprachiasmatic nucleus (SCN; Swaab et al. 1985, 1992). The SCN receives a strong input from the retina. It has been shown that exposure of AD patients to more intense light improves sleep-wake rhythms in clinical trials of these patients (Hozumi et al. 1990; Okawa et al. 1991; Mishima et al. 1994; Satlin et al. 1992). These clinical observations agree with the experimental data showing that degenerative changes in the SCN of aged rats, such as decreased amplitude of sleep-wake parameters and decreased numbers of vasopressin-expressing neurons in the SCN, could be counteracted by stimulation of the SCN by light (Witting et al. 1993; Lucassen et al. 1995).

The best way to prove that decreased metabolic activity indeed plays a major role in the development of dementia is of course to show in the future that reversing decreased neuronal activity would lead to considerable improvement of cognitive functions. The observation that glucose administration enhances memory in patients with probable AD (Manning et al. 1993) indicates that the focus on metabolic stimulation of neurons might be a fruitful strategy.

Summary and Conclusions

1. The neuropathological hallmarks of AD, i.e. amorphous plaques, NPs, pretangles, NFT and cell death are not part of a pathogenetic cascade but are basically independent phenomena.
2. Pretangles can occur in neurons in which the metabolic rate is not altered. However, in brain areas where classical AD changes, i.e., NPs and NFT, are present, such as the CA1 area of the hippocampus, the nucleus basalis of Meynert and the tuberomamillary nucleus, a decreased metabolic rate is found.
3. Decreased metabolic rate also seems to be an independent phenomenon in AD. It is not induced by the presence of pretangles, NFT or NPs.
4. Decreased metabolic rate is considered to be crucial and an early hallmark of AD. In theory, decreased metabolic rate may be reversible. It is, therefore,

attractive to direct the development of therapeutic strategies towards restimulation of neuronal metabolic rate to improve cognition and other symptoms in AD.

Acknowledgments

This paper was written as part of the AMSTEL project and supported by the "Stimuleringsprogramma Gezondsheidsonderzoek" (SGO) of the Ministry of Education and Science in The Netherlands (J. A. P. Van de Nes). P. J. Lucassen was supported by NWO (Grant no 100-007) and A. Salehi by a grant from the Shahid Beheshti University of Medical Sciences, Teheran, Iran. Brain material used was obtained from the Netherlands Brain Bank in the Netherlands Institute for Brain Research, Amsterdam (coordinator: Dr. R. Ravid).

References

Akiyama H, Tago H, Itagaki S, McGeer PL (1990) Occurrence of diffuse amyloid deposits in the presubicular parvopyramidal layers in Alzheimer's disease. Acta Neuropath (Berl) 79: 537–544

Allen SJ, Dawbarn D, Wilcock GK (1988) Morphometric immunochemical analysis of neurons in the nucleus basalis of Meynert in Alzheimer's disease. Brain Res 454: 275–281

Andrä K, Abramowski D, Duke M, Probst A, Wiederhold K-H, Bürki K, Goedert M, Sommer B, Staufenbiel M (1996) Expression of APP in transgenic mice: a comparison of neuron-specific promotors. Neurobiol Aging 17: 183–190

Armstrong RA, Myers D, Smith CUM (1992) Alzheimer's disease: Are cellular neurofibrillary tangles linked to beta/A4 formation at the projection sites? Neurosci Res Comm 11: 171–177

Armstrong RA, Myers D, Smith CUM (1993) The spatial patterns of plaques and tangles in Alzheimer's disease do not support the 'cascade hypothesis.' Dementia 4: 16–20

Arnold SE, Hyman BT, Flory J, Damasio AR, Van Hoesen GW (1991) The topographical and neuroanatomical distribution of neurofibrillary tangles and neuritic plaques in the cerebral cortex of patients with Alzheimer' disease. Cereb Cortex 1: 103–106

Arriagada PV, Growdon JH, Hedly-Whyte ET, Hyman BT (1992) Neurofibrillary tangles but not senile plaques parallel duration and severity of Alzheimer's disease. Neurology 42: 631–639

Bancher C, Brunner C, Lassmann H, Budka H, Jellinger K, Wiche G, Grundke-Iqbal I, Iqbal K, Wisnieswski HM (1989) Accumulation of abnormally phosphorylated tau precedes the formation of neurofibrillary tangles in Alzheimer's disease. Brain Res 477: 90–99

Barcikowska M, Wisniewski HM, Bancher C, Grundke-Iqbal I (1989) About the presence of paired helical filaments in dystrophic neurites participating in the plaque formation. Acta Neuropath (Berl) 78: 225–231

Bondareff W, Mountjoy CQ, Roth M (1982) Loss of neurons of origin of the adrenergic projection to cerebral cortex (nucleus locus ceruleus) in senile dementia. Neurology 32: 164–168

Bouras C, Hof PR, Giannakopoulos P, Michel J-P, Morrison JH (1994) Regional distribution of neurofibrillary tangles and senile plaques in the cerebral cortex of elderly patients: a quantitative evaluation of a one-year autopsy population from a geriatric hospital. Cereb Cortex 4: 138–150

Bowen DM, Smith CB, White P, Flack RHA, Carrasco LH, Gedye JL, Davidson AN (1977) Chemical pathology of the organic dementias. II. Quantiative estimation of cellular changes in post-mortem brains. Brain 100: 427–453

Braak E, Braak H, Mandelkow E-M (1994) A sequence of cytoskeletal changes related to the formation of neurofibrillary tangles and neuropil threads. Acta Neuropathol (Berl) 8733: 554–567

Braak H, Braak E (1990) Alzheimer's disease: Striatal amyloid deposits and neurofibrillary changes. J Neuropath Exp Neurol 49: 15–224

Braak H, Braak E (1991) Neuropathological staging of Alzheimer-related changes. Acta Neuropath (Berl) 82: 239–259

Braak H, Braak E, Grundke-Iqbal I (1986) Occurrence of neuropil threads in the senile human brain and in Alzheimer's disease: a third location of paired helical filaments outside of neurofibrillary tangles and neuritic plaques. Neurosci Lett 65: 351–355

Braak H, Braak E, Bohl J, Lang W (1989a) Alzheimer's disease: amyloid plaques in the cerebellum. J Neurol Sci 93: 277–287

Braak H, Braak E, Ohm T, Bohl J (1989b) Alzheimer's disease: mismatch between amyloid plaques and neuritic plaques. Neurosci Lett 103: 24–28

Chan-Palay V, Asan E (1989) Alterations in catecholamine neurons of the locus coeruleus in senile dementia of the Alzheimer type and in Parkinson's disease with and without dementia and depression. J Comp Neurol 287: 373–392

Crystal H, Dickson D, Fuld P, Masur D, Scott R, Mehler M, Masdeu J, Kawas C, Aronson M, Wolfson L (1988) Clinico-pathologic studies in dementia: nondemented subjects with pathologically confirmed Alzheimer's disease. Neurology 38: 1682–1687

Davies L, Wolska B, Hilbich C, Multhaup G, Martins R, Simms G, Beyreuther K, Masters CL (1988) A4 amyloid protein deposition and the diagnosis of Alzheimer's disease: Prevalence in aged brains determined by immunocytochemistry compared with conventional neuropathologic techniques. Neurology 38: 1688–1693

De Lacalle S, Iraizoz I, Ma Conzalo L (1992) Differential changes in cell size and number in topographic subdivisions of human basal nucleus in normal aging. Neuroscience 43: 445–456

Delaere P, Duyckaerts C, Masters CL, Beyreuther K (1990) Large amounts of neocortical beta/A4 deposits without neuritic plaques nor tangles in psychometrically assessed, non-demented person. Neurosci Lett 116: 87–93

Delaere P, Yi HE, Fayet G, Duyckaerts C, Hauw J-J (1993) Beta/A4 deposits are constant in the brain of the oldest old: an immunohistochemical study of 20 French centenarians. Neurobiol Aging 14: 191–194

Dickson DW, Farlo J, Davies P, Crystal H, Fuld P, Yen S-HC (1988) Alzheimer's disease: A double-labelling immunohistochemical study of senile plaques. Am J Path 132: 86–101

Doebler JA, Markesberry WR, Anthony A, Davies P, Scheff SW, Rhoads RE (1988) Neuronal RNA in relation to Alz-50 immunoreactivity in Alzheimer's disease. Ann Neurol 23: 20–24

Duyckaerts C, Hauw J-J, Bastenaire F, Piette F, Poulain C, Rainsard V, Javoy-Agid F, Berthaux P (1986) Laminar distribution of neocortical senile plaques in senile dementia of the Alzheimer type. Acta Neuropathol (Berl) 70: 249–256

Duyckaerts C, Delaere P, Poulain V, Brion J-P, Hauw J-J (1988) Does amyloid precede paired helical filaments in the senile plaque? A study of 125 cases with graded intellectual status in aging and Alzheimer disease. Neurosci Lett 91: 354–359

Emre M, Geula C, Ransil BJ, Mesulam M-M (1992) The acute neurotoxicity and effects on cholinergic axons of interacerebrally injected β amyloid in the rat brain. Neurobiol Aging 13: 553–560

Foster NL, Chase TN, Mansi L, Brooks R, Fedio P, Patronas NJ, Di Chiro G (1984) Cortical abnormalities in Alzheimer's disease. Ann Neurol 16: 649–654

Games D, Adams D, Alessandrini R, Barbour R, Berthelette P, Blackwell C, Carr T, Clemens J, Donaldson T, Gillespie F, Guido T, Hagopian S, Johnson-Wood K, Khan K, Lee M, Liebowitz P, Lieberburg I, Little S, Masliah E, McConlogue L, Montoya-Zavala M, Mucke L, Paganini L, Penniman E, Powe M, Schenk D, Seubert P, Snyder B, Soriano F, Tan H, Vital J, Wadsworth S, Wolozin B, Zhao J (1995) Alzheimer type neuropathology in transgenic mice overexpressing V717F β-amyloid precursor protein. Nature 373: 523–527

German DC, Manaye KF, White III CL, Woodward DJ, McIntire DD, Smith WK, Kalaria RN, Mann DMA (1992) Disease-specific patterns of locus coeruleus cell loss. Ann Neurol 32: 667–676

Gertz HJ, Schoknecht G, Krüger H, Cervos-Navarro J (1989) Stability of cell size and nucleolar size in tangle-bearing neurons of hippocampus in Alzheimer's disease. Brain Res 487: 373–375

Goudsmit E, Hofman MA, Fliers E, Swaab DF (1990) The supraoptic and paraventricular nuclei of the human hypohalamus in relation to sex, age and Alzheimer's disease. Neurobiol Aging 11: 529–536

Greenberg BD, Savage MJ, Howland DS, Ali SM, Siedlak SL, Perry G, Siman R, Scott RW (1996) APP transgenesis: approaches toward the development of animal models for Alzheimer disease neuropathology. Neurobiol Aging 17: 153–171

Guillemette JG, Wong L, Crapper McLachlan DR, Lewis PN (1986) Characterization of messenger RNA from the cerebral cortex of control and Alzheimer-afflicted brain. J Neurochem 47: 987–997

Haxby JV, Grady CL, Koss E, Horwitz B, Schapiro M, Friedland RP, Rapoport SI (1988) heterogeneous anterior-posterior metabolic patterns in dementia of the Alzheimer type. Neurology 38: 1853–1863

Hoogendijk WJG, Pool CW, Troost D, Van Zwieten EJ, Swaab DF (1995) Image-analyzer-assisted morphometry of the locus coeruleus in Alzheimer's disease, Parkinson's disease and amyotrophic lateral sclerosis. Brain 118: 131–143

Hoyer S, Oesterreich K, Wagner O (1988) Glucose metabolism as the site of the primary abnormality in early-onset dementia of Alzheimer type? J Neurol 235: 143–148

Hozumi S, Okawa M, Mishima K, Hishikawa Y, Hori H, Takahashi K (1990) Phototherapy for elderly patients with dementia and sleep-wake rhythm disorders – a comparison between morning and evening light exposure. Japan J Psych Neurol 44: 813–814

Hyman BT, Van Hoesen GW, Kromer LJ, Damasio AR (1986) Perforant pathway changes and the memory impairment of Alzheimer's disease. Ann Neurol 20: 472–481

Joachim CL, Morris JH, Selkoe DJ (1989) Diffuse senile plaques occur commonly in the cerebellum in Alzheimer's disease. Am J Pathol 135: 309–319

Jucker M, Walker LC, Martin LJ, Kitt CA, Kleinman HK, Ingram DK, Price DL (1992) Age-associated inclusions in normal and transgenic mouse brain. Science 255: 1443–1445

Kalus P, Braak H, Braak ER, Bohl J (1989) The presubicular region in Alzheimer's disease: topography of amyloid deposits and neurofibrillary changes. Brain Res 494: 198–203

Kammesheidt A, Boyce FM, Spanoyannis AF, Cummings BJ, Ortegon M, Cotman C, Vaught JL, Neve RL (1992) Deposition of beta/A4 immunoreactivity and neuronal pathology in transgenic mice expressing the carboxy terminal fragment of the Alzheimer amyloid precursor in the brain. Proc Natl Acad Sci USA 89: 10857–10861

Kang J, Lemaire H-G, Unterbeck A, Salbaum JM, Masters CL, Grzeschik K-H, Multhaup G, Beyreuther K, Muller-Hill B (1987) The precursor of Alzheimer's disease amyloid A4 protein resembles a cell-surface receptor. Nature 325: 733–736

Katzman R, Terry RD, DeTeresa R, Brown T, Davies P, Fuld P, Renbing X, Peck A (1987) Clinical, pathological and neurochemical changes in dementia, a subgroup with preserved mental status and numerous neocortical plaques. Ann Neurol 23: 138–144

Kawabata S, Higgins GA, Gordon JW (1991) Amyloid plaques, neurofibrillary tangles and neuronal loss in brains of transgenic mice overexpressing a C-terminal fragment of human amyloid precursor protein. Nature 354: 476–478

Kowall NW, Kosik KS (1987) Axonal disruption and aberrant localization of tau protein characterize the neuropil pathology of Alzheimer's disease. Ann Neurol 22: 639–643

Kowall N, McKee AC, Yankner BA, Beal MF (1992) In vivo neurotoxicity of beta-amyloid [$\beta(1-40)$] and $\beta(25-35)$ fragment. Neurobiol Aging 13: 537–542

Kremer HPH (1992) The hypothalamic lateral tuberal nucleus: normal anatomy and changes in neurological diseases. In: Swaab DF, Hofman MA, Mirmiran M, Ravid R, van Leeuwen FW (eds) The human hypothalamus in health and disease. Progress in Brain Research, Vol 93. Elsevier, Amsterdam, pp 249–263

Kremer HPH, Swaab DF, Bots GThAM, Fisser B, Ravid R, Roos RAC (1991) The hypothalamic lateral tuberal nucleus in Alzheimer's disease. Ann Neurol 29: 279–284

Kumar A, Newberg A, Alavi A, Berlin J, Smith R, Reivich M (1993) Regional cerebral glucose metabolism in late-life Alzheimer disease: a preliminary positron emission. Proc Natl Acad Sci USA 90: 7019–7023

Lee VM, Balin BJ, Otvos L Jr, Trojanowski JQ (1991) A68: a major subunit of paired helical filaments and forms of normal Tau. Science 251: 675–678

Levy-Lahad E, Wijsman EM, Nemens E, Anderson L, Goddard KAB, Weber E, Bird TD, Schellenberg GD (1995) A fimilial Alzheimer's disease locus on chromosome 1. Science 269: 970–973

Lowes-Hummel P, Gertz H-J, Ferszt R, Cervos-Navarro J (1989) The basal nucleus of Meynert revised: the nerve cell number decreases with age. Arch Gerontol Geriatrics 8: 21–27

Lucassen PJ, Salehi A, Pool CW, Gonatas NK, Swaab DF (1994) Activation of vasopressin neurons in aging and Alzheimer's disease. J Neuroendocr 6: 673–679

Lucassen PJ, Hofman MA, Swaab DF (1995) Increased light intensity prevents the age related loss of vasopressin-expressing neurons in the rat suprachiasmatic nucleus. Brain Res 693: 261–266

Malherbe P, Richards JG, Martin JR, Bluethmann H, Maggio J, Huber G (1996) Lack of β-amyloidosis in transgenic mice expressing low levels of familial Alzheimer's disease missense mutations. Neurobiol Aging 17: 205–214

Mann DMA, Jones D (1990) Deposition of amyloid (A4) protein within the brains of persons with dementing disorders other than Alzheimer's disease and Down's syndrome. Neurosci Lett 109: 68–75

Mann DMA, Neary D, Yates PO, Lincoln J, Snowden JS, Stanworth P (1981) Alterations in protein synthetic capability of nerve cells in Alzheimer's disease. J Neurosurg Psychiat 44: 97–102

Mann DMA, Yates PO, Marcyniuk B (1984) Changes in nerve cells of the nucleus basalis of Meynert in Alzheimer's disease and their relationship to ageing and to the accumulation of lipofuscin pigment. Mechan Ageing Dev 25: 189–204

Mann DMA, Yates PO, Marcyniuk B (1985) Some morphometric observations in the cerebral cortex and hippocampus in presenile Alzheimer's disease, senile dementia of Alzheimer type and Down's syndrome in middle age. J Neurol Sci 69: 139–159

Manning CA, Ragozzino ME, Gold PE (1993) Glucose enhancement of memory in patients with probable senile dementia of the Alzheimer's type. Neurobiol Aging 14: 523–528

Mantione JR, Kleppner SR, Miyazono M, Wertkin AM, Lee VM-Y, Trojanowski JQ (1995) Human neurons that constitutively secrete Aβ do not induce Alzheimer's disease pathology following transplantation and long-term survival in the rodent brain. Brain Res 671: 333–337

Marcus DL, de Leon MJ, Goldman J, Logan J, Christman DR, Wolf AP, Fowler JS, Hunter K, Tsai J, Pearson J, Freedman ML (1989) Altered glucose metabolism in microvessels from patients with Alzheimer's disease. Ann Neurol 26: 91–94

Masters CL, Simms G, Weinman NA, Multhaup G, McDonald BL, Beyreuther K (1985) Amyloid core plaque protein in Alzheimer disease and Down syndrome. Proc Natl Acad Sci USA 82: 4245–4249

McKee AC, Kosik KS, Nowall NW (1991) Neuritic pathology and dementia in Alzheimer's disease. Ann Neurol 30: 156–165

Meier-Ruge W, Bertoni-Freddari C, Iwangoff P (1994) Changes in brain glucose metabolism as a key to the pathogenesis of Alzheimer's disease. Gerontology 40: 246–252

Meneilly GS, Hill A (1993) Alterations in glucose metabolism in patients with Alzheimer's disease. J Am Geriatr Soc 41: 710–714

Mesulam MM, Mufson MJ, Levey AI, Wainer BH (1983) Cholinergic innervation of cortex by the basal forebrain: Cytochemistry and cortical connections of the septal area, diagonal band nuclei, nucleus basalis (substantia innominata), and hypothalamus in the rhesus monkey. J Comp Neurol 214: 170–197

Mielke R, Herholz K, Grond M, Kessler J, Heiss WD (1994) Clinical deterioration in probable Alzheimer's disease correlates with progressive metabolic impairment of association areas. Dementia 5: 36–41

Mishima K, Okawa M, Hishikawa Y, Hozumi S, Hori H, Takahashi K (1994) Morning bright light therapy for sleep and behavior disorders in elderly patients with dementia. Acta Psychiatr Scand 89: 1–7

Morrison JH, Foote SL, O'Conner D, Bloom FE (1982) Laminar, tangential and regional organization of the noradrenergic innervation of monkey cortex: dopamine-beta-hydroxylase immunohistochemistry. Brain Res Bull 9: 309–319

Mukaetova-Ladinska EB, Harrington CR, Roth M, Wischik CM (1993) Biochemical and anatomical redistribution of tau protein in Alzheimer's disease. Am J Pathol 143: 565–578

Mullan M, Houlden H, Windelspect M, Fidani L, Lombardi C, Diaz P, Rossor M, Crook R, Hardy J, Duff K, Crawford F (1992) A locus for familial early-onset Alzheimer's disease on the long arm of chromosome 14, proximal to the alpha 1-anti-chymotrypsin gene. Nature Genet 2: 340–342

Nakamura S, Takemura M, Ohnishi K, Suenaga T, Nishimura M, Akiguchi I, Kimura J, Kimura T (1993) Loss of large neurons and occurrence of neurofibrillary tangles in the tuberomammillary nucleus of patients with Alzheimer's disease. Neurosci Lett 151: 196–199

Ogomori K, Kitamoto T, Tateishi J, Sato Y, Suetsugu M, Abe M (1990) Beta-protein amyloid is widely distributed in the central nervous system of patients with Alzheimer's disease. Am J Pathol 134: 243–251

Okawa M, Mishima K, Hishikawa Y, Hozumi S, Hori H, Takahashi K (1991) Circadian rhythm disorders in sleep-waking and body temperature in elderly patients with dementia and their treatment. Sleep 14: 478–485

Pearson RCA, Gatter KC, Powell TPS (1983) Retrograde cell degeneration in the basal nucleus in monkey and man. Brain Res 261: 321–326

Pearson RCA, Esiri MM, Hiorns RW, Wilcock GK, Powell TPS (1985) Anatomical correlates of the distribution of the pathological changes in the neocortex in Alzheimer's disease. Proc Natl Acad Sci USA 82: 4531–4534

Peppard RF, Martin WRW, Carr GD, Grochowski E, Schulzer M, Guttman M, McGeer PL, Phillips AG, Tsui JKC, Calne DB (1992) Cerebral glucose metabolism in Parkinson's disease with and without dementia. Arch Neurol 49: 1262–1268

Pericak-Vance MA, Bebout JL, Gaskell Jr PC, Yamaoka LH, Hung WY, Alberts MJ, Walker AP, Bartlett RJ, Haynes CA, Welsh KA, Earl NL, Heyman A, Clark CM, Roses AD (1991) Linkage studies in familial Alzheimer disease: evidence for chromosome 19 linkage. Am J Human Genet 48: 1034–1050

Powell RR (1974) Psychological effects of exercise therapy in institutionalized geriatric mental patients. J Gerontol 29: 157–164

Quon D, Wang Y, Catalano R, Scardina JM, Murakami K, Cordell B (1991) Formation of beta-amyloid protein deposits in brains of transgenic mice. Nature 352: 239–241

Regeur L, Jensen GB, Pakkenberg H, Evans SM, Pakkenberg B (1994) No global neocortical nerve cell loss in brains from senile dementia of Alzheimer's type. Neurobiol Aging 15: 347–352

Reiman EM, Caselli RJ, Yun LS, Chen K, Bandy D, Minoshima S, Thibodeau SN, Osborne D (1996) Preclinical evidence of Alzheimer's disease in persons homozygous for the ε4 allele for apolipoprotein E. N Engl J Med 334: 752–758

Rinne JO, Paljarvi L, Rinne UK (1987) Neuronal size and density in the nucleus basalis of Meynert in Alzheimer's disease. J Neurol Sci 79: 67–76

Roberts GW, Allsop D, Bruton CJ (1990) The occult aftermath of boxing. J Neurol Neurosurg Psychiatr 24: 173–182

Rogers J, Morrison JH (1985) Quantitative morphology and regional and laminar distributions of senile plaques in Alzheimer's disease. J Neurosci 5: 2801–2808

Rossor MN (1993) Molecular pathology of Alzheimer's disease. J Neurol Neurosurg Psychiatr 56: 583–586

Rozemuller JM, Eikelenboom P, Stam FC, Beyreuther K, Masters CL (1989) A4 protein in Alzheimer's disease; primary and secondary cellular events in extracellular amyloid deposition. J Neuropath Exp Neurol 48: 674–691

Rudelli RD, Ambler MW, Wisniewski HM (1984) Morphology and distribution of Alzheimer neuritic (senile) and amyloid plaques in striatum and diencephalon. Acta Neuropathol (Berl) 64: 273–381

Sajdel-Sulkowska EM, Marotta CA (1984) Alzheimer's disease brain: Alterations in RNA levels and in a ribonuclease-inhibitor complex. Science 225: 947–949

Salehi A, Lucassen PJ, Pool CW, Gonatas NK, Ravid R, Swaab DF (1994) Decreased neuronal activity in the nucleus basalis of Alzheimer's disease as suggested by the size of the Golgi apparatus. Neuroscience 59: 871–880

Salehi A, Van de Nes JAP, Hofman MA, Gonatas NK, Swaab DF (1995a) Early cytoskeletal changes as shown by Alz-50 are not accompanied by decreased neuronal activity. Brain Res 678: 29–39

Salehi A, Ravid R, Gonatas NK, Swaab DF (1995b) Decreased activity of hippocampal neurons in Alzheimer's disease is not related to the presence of neurofibrillary tangles. J Neuropathol Exp Neurol 54: 704–709

Salehi A, Heyn S, Gonatas NK, Swaab DF (1995c) Decreased protein synthetic activity of the hypothalamic tuberomammillary nucleus in Alzheimer's disease as suggested by a smaller Golgi apparatus. Neurosci Lett 193: 29–32

Salmon E, Gregoire MC, Delfiore G, Lemaire C, Degueldre C, Franck G, Comar D (1996) Combined study of cerebral glucose metabolism and [^{11}C]methionine accumulation in probable Alzheimer's disease using positron emission tomography. J Cerebr Blood Flow Metab 16: 399–408

Satlin A, Volicer L, Ross V, Herz L, Campbell S (1992) Bright light treatment of behavioral and sleep disturbances in patients with Alzheimer's disease. Am J Psychiatr 149: 1028–1032

Schellenberg GD, Bird TD, Wijsman EM, Orr HT, Anderson L, Nemens E, White JA, Bonnycastle L, Weber JL, Alonso ME, Potter H, Heston LL, Martin GM (1992) Genetic linkage evidence for a familial Alzheimer's disease locus on chromosome 14. Science 258: 668–671

Scherder E, Bouma A, Steen L, Swaab D (1995a) Peripheral nerve stimulation in Alzheimer's disease. A meta-analysis. Alz Res 1: 183–184

Scherder EJA, Bouma A, Steen AM (1995b) Effects of short-term transcutaneous electrical nerve stimulation on memory and affective behaviour in patients with probable Alzheimer's disease. Behav Brain Res 67: 211–219

Scherder EJA, Bouma A, Steen AM (1995c) Effects of simultaneously applied short-term transcutaneous electrical nerve stimulation and tactile stimulation on memory and affective behaviour of patients with probable Alzheimer's disease. Behav Neurol 8: 3–13

Selkoe DJ (1994) Alzheimer's disease: a central role for amyloid. J Neuropathol Exp Neurol 53: 438–447

Simonian NA, Hyman BT (1993) Functional alterations in Alzheimer's disease: diminution cytochrome oxidase in the hippocampal formation. J Neuropathol Exp Neurol 52: 580–585

Simonian NA, Hyman BT (1994) Functional alterations in Alzheimer's disease: selective mitochondrial-encoded cytochrome oxidase mRNA in the formation. J Neuropathol Exp Neurol 53: 508–512

Standaert DG, Lee VM-Y, Greenberg BD, Lowery DE, Trojanowski JQ (1991) Molecular features of hypothalamic plaques in Alzheimer's disease. Am J Pathol 139: 681–691

St George-Hyslop PH, Haines J, Rogaev E, Mortilla M, Vaula G, Pericak-Vance M, Foncin J-F, Montesi M, Bruni A, Sorbi S, Rainero I, Pinessi L, Pollen D, Polinsky R, Nee L, Kennedy J, Macciardi, Rogaeva E, Liang Y, Alexandrova N, Lukiw W, Schlumpf K, Tanzi R, Tsuda T, Farrer L, Cantu J-M, Duara R, Amaducci L, Bergamini L, Gusella J, Roses A, Crapper McLachlan D (1992) Genetic evidence for a novel familial Alzheimer's disease locus on chromosome 14. Nature Genet 2: 330–334

Strittmatter WJ, Saunders AM, Schmechel D, Pericak-Vance M, Enghild J, Salvesen GS, Roses AD (1993a) Apolipoprotein E: high-avidity binding to beta-amyloid and increased frequency of type allele in late-onset familial Alzheimer disease. Proc Natl Acad Sci USA 90: 1977–1981

Strittmatter WJ, Weisgraber KH, Huang DY, Dong L-M, Salvesen GS, Pericak-Vance M, Schmechel D, Saunders AM, Goldgaber D, Roses AD (1993b) Binding of human apolipoprotein E to synthetic amyloid beta peptide; isoform specific effects and implications for late-onset Alzheimer disease. Proc Natl Acad Sci USA 90: 8098–8102

Suzuki K, Katzman R, Korey SR (1965) Chemical studies on Alzheimer's disease. J Neuropathol Exp Neurol 24: 211–214

Swaab DF (1991) Brain aging and Alzheimer's disease: "wear and tear" versus "use it or lose it". Neurobiol Aging 12: 317–324

Swaab DF, Hofman MA (1988) Sexual differentiation of the human hypothalamus: ontogeny of the sexually dimorphic nucleus of the preoptic area. Dev Brain Res 44: 314–318

Swaab DF, Fliers E, Partiman T (1985) The suprachiasmatic nucleus of the human brain in relation to sex, age and dementia. Brain Res 342: 37–44

Swaab DF, Grundke-Iqbal I, Iqbal K, Kremer HPH, Ravid R, Van de Nes JAP (1992) Tau and ubiquitin in the human hypothalamus in aging and Alzheimer's disease. Brain Res 590: 239–249

Swerdlow R, Marcus DL, Landman J, Kooby D, Frey W, Freedman ML (1994) Brain glucose metabolism in Alzheimer's disease. Am J Med Sci 308: 141–144

Tagliavini F, Giaccone G, Verga L, Ghiso J, Frangione B, Bugiani O (1991) Alzheimer patients: preamyloid deposits are immunoreactive with antibodies to extracellular domains of the amyloid precursor protein. Neurosci Lett 128: 117–120

Tate B, Aboody-Guterman KS, Morris AM, Walcott EC, Majocha RE, Marotta CA (1992) Disruption of circadian regulation by brain grafts that overexpresses Alzheimer β/A4 amyloid. Proc Natl Acad Sci USA 89: 7090–7094

Taylor GR, Carter GI, Crow TJ, Johnson JA, Fairbairn AF (1986) Recovery and measurement of specific RNA species from tissue: a general reduction in Alzheimer's disease detected by hybridization. Exp Mol Pathol 44: 111–116

Terry RD, Hansen LA, DeTeresa R, Davies P, Tobias H, Katzman R (1987) Senile dementia of the Alzheimer type without neocortical neurfibrillary tangles. J Neuropath Exp Neurol 46: 262–268

Van Broeckhoven C, Backhovens H, Cruts M, De Winter G, Bruyland M, Cras P, Martin J-J (1992) Mapping of a gene predisposing to early-onset Alzheimer's disease to chromosome 14q24.3. Nature Genet 2: 335–339

Van de Nes JAP, Kamphorst W, Ravid R, Swaab DF (1993) The distribution of Alz-50 immunoreactivity in the hypothalamus and adjoining areas of Alzheimer's disease patients. Brain 116: 103–115

Van de Nes JAP, Kamphorst W, Swaab DF (1994) Arguments for and against the primary amyloid local induction hypothesis of the pathogenesis of Alzheimer's disease. Ann Psychiat 4: 95–111

Van der Woude PF, Goudsmit E, Wierda M, Purba JS, Hofman MA, Bogte H, Swaab DF (1995) No vasopressin cell loss in the human paraventricular and supraoptic nucleus during aging and in Alzheimer's disease. Neurobiol Aging 16: 11–18

West MJ, Coleman PD, Flood DG, Tronsoco JC (1994) Differences in the pattern of hippocampal neuronal loss in normal ageing and Alzheimer's disease. Lancet 344: 769–772

Whitehouse PJ, Price DL, Struble RG, Clark AW, Coyle JT, Delong MR (1982) Alzheimer's disease and senile dementia: loss of neurons in the basal forebrain. Science 215: 1237–1239

Whitehouse PJ, Hedreen JC, White CL, Clark AW, Price DL (1983) Neuronal loss in the basal forebrain cholinergic system is more marked in Alzheimer's disease than in senile dementia of the Alzheimer type. Ann Neurol 14: 149

Wirak DO, Bayney R, Ramabhadran TV, Fracasso RP, Hart JT, Hauer PE, Hsiau P, Pekar SK, Scangos GA, Trapp BD, Unterbeck AJ (1991) Deposits of amyloid protein in the central nervous system of transgenic mice. Science 253: 1–2

Wisniewski HM, Wegiel J (1995) The neuropathology of Alzheimer's disease. Neuroimaging Clin N Am 5 (1): 45–57

Witting W, Kwa IH, Eikelenboom P, Mirmiran M, Swaab DF (1990) Alterations in the circadian rest-activity rhythm in aging and Alzheimer's disease. Biol Psychiatr 27: 563–572

Witting W, Mirmiran M, Bos NP, Swaab DF (1993) Effect of light intensity on diurnal sleep-wake distribution in young and old rats. Brain Res Bull 30: 157–162

Neurodegenerative Alzheimer-like Pathology in PDAPP 717V → F Transgenic Mice

D. Games[*], *E. Masliah, M. Lee, K. Johnson-Wood, and D. Schenk*

Summary

Predominant pathological hallmarks of Alzheimer's (AD) include the region-specific deposition of β amyloid (Aβ) plaques, vascular amyloidosis, and a number of distinct neurodegenerative changes. These involve the formation of dystrophic neurites and neuritic plaques, cytoskeletal alterations, and synaptic and neuronal loss. Astrocytosis and microgliosis are also evident in affected brain regions. Transgenic (tg) mice overexpressing a mutant form of the β-amyloid precursor protein (APP 717 V → F) develop several of these pathologies in an age- and region-dependent manner similar to AD. Aβ plaques in the transgenic mouse share many of the tinctorial and immunohistochemical properties of AD plaques, including the relative distribution of $A\beta_{X-40}$ and $A\beta_{X-42}$ isoforms and the presence of other plaque-associated proteins. Initial findings using immunoassays specific to unique forms of APP and Aβ suggest that APP levels do not dramatically change with increasing age in the mouse brains, and that region-specific variations in APP metabolism, local factors or deficits in distinct populations of neurons account for the deposition of brain Aβ. The PDAPP mouse is a relevant and efficient model system to identify mechanistic properties of the disease process and offers novel opportunities to test potential therapeutics.

Introduction

Several pathologic characteristics of Alzheimer's disease (AD) develop in a line of transgenic mice (Games et al. 1995) overexpressing a human amyloid precursor (hAPP) mutation, hAPP717V → F, associated with familial AD (Murrell et al. 1991). The transgenic construct employs a platelet-derived growth factor (PDGF) promoter to drive a minigene that allows alternative splice site selection (Rockenstein et al. 1995) of the precursor protein, resulting in differential production of the hAPP isoforms hAPP770, hAPP751 and hAPP695. Transgenic mRNA is increased 4- to 6-fold over non-transgenic mouse APP, expressed against a mixed background strain of Swiss Webster, DBA/2 and C57BL/6. The

---

[*] Athena Neurosciences 800 Gateway Blvd. South San Francisco, California 94002

B. T. Hyman / C. Duyckaerts / Y. Christen (Eds.)
Connections, Cognition, and Alzheimer's Disease
© Springer-Verlag Berlin Heidelberg 1997

brains of these mice develop β amyloid (Aβ) deposition, neuritic dystrophy, synaptic and dendritic loss, gliosis and cytoskeletal alterations. Notably, the lesions affect brain regions that are associated with early and extensive damage in AD. These animals, called PDAPP mice, provide novel opportunities for obtaining a mechanistic understanding of the disease process and afford an efficient and convenient system for testing potential therapeutics. This discussion will compare PDAPP mouse brain pathology with that of AD, and address several issues pertaining to its use as a system for the study of brain region vulnerability and differential APP processing in the disease.

PDAPP and AD Pathology

The predominant neuropathologies in AD include the following features: 1) Aβ deposition into senile plaques (SP) and the formation of amyloidotic vasculature (Wisniewski et al. 1989, Selkoe 1991; Yamaguchi et al. 1988) 2) neurodegeneration (Hyman et al. 1990; Braak and Braak 1995; Masliah et al. 1993a) and 3) astrocytosis, microgliosis and the deposition of acute phase proteins (Frederickson 1992; Snow et al. 1988; Mann 1994; Asian and Davis 1994). Pathological characterization of PDAPP mice brains revealed that many of these lesions occurred in an age- and region-dependent manner remarkably similar to AD.

Aβ Deposition

The cerebral deposition of Aβ, called senile plaques, is recognized as an invariant and characteristic feature of AD pathology (Selkoe 1991). Aβ plaques are generally classified as belonging to either the "diffuse" or "compacted" form (Yamaguchi et al. 1988). The former consists of fine electrodense Aβ aggregates, lacks substantial fibrillar amyloid and a concentrated core, and is not associated with extensive neuritic alterations or gliosis. In contrast, the compacted plaques contain dense and fibrillar Aβ, have a defined boundary, may have a well-defined core, and are usually associated with dystrophic neurites, astrocytosis and microgliosis. In addition to these numerous degenerative changes in the neuropil, compacted plaques are associated with a number of extracellular proteins, including complement and acute phase reactants (McGeer et al. 1987; Abraham et al. 1988) as well as lysosomal hydrolases (Cataldo et al. 1994). They are also labeled by a number of tinctorial and histochemical staining methods, such as thioflavin S, Congo red, and silver stains. Aβ plaques typically do not develop throughout the brain, but are formed predominantly in the association and limbic cortices. This dramatic selective vulnerability implies that region-specific proteins, metabolism or connectional factors induce Aβ formation and/or deposition; however, their precise roles remain largely undefined.

Prior to the development of the PDAPP transgenic mouse, the opportunity to examine these issues in an experimental fashion was extremely limited due to the lack of a suitable and convenient animal model. Although hippocampal and cortical Aβ deposition has been described in transgenic mice expressing hAPP751 (Higgins et al. 1994), canines (Cummings et al. 1993), polar bears (Selkoe et al. 1987) and primates (Price et al. 1991), the lengthy and variable time course of Aβ deposition and lack of advanced AD-type lesions prevented systematic and convenient approaches to study the disease process *in vivo*. These impediments, coupled with impractical and expensive husbandry issues, precluded opportunities for efficient experimental manipulation and compound testing for potential therapeutics.

The most informative studies of temporal progression of Aβ deposition involved examination of the brains of patients with Down Syndrome (DS, Mann 1988; Hyman 1992). The human amyloid precursor (hAPP) gene is located on chromosome 21 and is therefore overexpressed in this trisomy 21 disorder. DS patients develop classic AD-type pathology after their fifth decade of life (Mann 1989). Recent studies show that the earliest plaques in DS are of the diffuse type which is composed primarily of $A\beta_{X-42}$ (Iwatsubo et al. 1995; Lemere et al. 1996). $A\beta_{X-42}$ is also the predominant form in AD plaques, whereas $A\beta_{X-40}$ labels the core of a subset of plaques in both DS and AD (Fukumoto et al. 1996).

A relevant and practical transgenic mouse model of human disease should recapitulate the pathology associated with the disorder. Furthermore, both scientific and practical criteria require that the lesions be revealed by methods similar and pertinent to those used to define the pathology in humans.

None of the changes described below are found in age-matched, nontransgenic controls from the same background strain. The neuropathological changes in the transgenic mice have been confirmed in five generations of animals.

Amyloid deposition begins in the heterozygous PDAPP mouse brains between four and six months of age in the hippocampus and cingulate, retrosplenial and frontal cortices. The Aβ plaques are recognized by a panel of human-specific antibodies, including those specific for the Aβ species produced by cleavage at the β-secretase site, as well as the isoforms $A\beta_{X-42}$ and $A\beta_{X-40}$. Several of the early deposits are less than 20 μm in diameter and resemble the granular Aβ aggregates described in AD (Takahashi et al. 1990). Others form spidery, cotton-like deposits that resemble components of diffuse AD deposits. Site-specific antibodies indicate that the predominant species in the early plaques is $A\beta_{X-42}$.

Aβ production and deposition is accelerated in the affected areas between seven and nine months of age (Table 1). Plaque density is greatly enhanced during this period, resulting in increased numbers of both diffuse and compacted plaques (Fig. 1A) by one year of age in both heterozygotic and homozygotic mice. The deposits display classic AD-type morphology, ranging from the diffuse types of deposits to compacted plaques with well-defined cores. Similar to AD, many deposits are recognized by silver and thioflavin S staining methods, and several are birefringent under polarized light after being stained with Congo red. Immunolabeling with antibodies specific to $A\beta_{X-40}$ and $A\beta_{X-42}$ indicates that the majority of plaques are composed of the $A\beta_{X-42}$ isoform (Fig. 1B), whereas $A\beta_{X-40}$

Table 1. Levels of $A\beta_{total}$, $A\beta_{42}$, APPα/FL and APPβ present in the PDAPP mouse brain as a function of age and brain regions. Note the remarkable rise in Aβ levels seen in the hippocampus and cortex as the animals age. This is in contrast to relatively constant levels of other APP metabolites. No age-dependent increase in Aβ levels is seen in the thalamus, a brain region that exhibits no apparent amyloidosis and neurodegeneration in the PDAPP mouse

	Age			
	4 months	8 months	10 months	12 months
Aβtotal (pmol/gm)				
Hippocampus	38	586	2182	4000
Cortex	18.9	113	270	690
Thalamus	7	11	15	18
Aβ$_{42}$ (pmol/gm)				
Hippocampus				
Cortex	5.2	83	270	668
Thalamus				
APPα/FL (pmol/gm)				
Hippocampus	703	660	650	678
Cortex	446	494	491	444
Thalamus	637	671	693	708
APPβ (pmol/gm)				
Hippocampus	198	174	159	134
Cortex	126	102	87	76
Thalamus	70	57	46	35

labels the core of a subset of plaques (Fig. 1C). $A\beta_{X-40}$ accumulates in regions affected by early amyloid deposition, indicating that the more mature plaques contain this isoform. $A\beta_{X-40}$ is also associated with amyloidotic blood vessels in the meninges and subpial matter, also resembling those formed in AD. Ultra-structural studies (Masliah et al., submitted for publication; Fig. 4) show that the PDAPP deposits consist of extracellular amyloid fibrils that are organized into amyloid "stars" that typify Aβ plaques in AD; however, they are somewhat dense by comparison. A number of the plaque-associated proteins, including the proteoglycan glypican, as well as apolipoprotein E and the lysosomal enzyme cathepsin D are also associated with PDAPP Aβ deposits.

The outer molecular layer (OML) of the hippocampal dentate gyrus develops extensive Aβ deposition by one year of age in PDAPP brains (Fig. 1D), and is associated with pronounced Aβ deposition in AD. This region receives the terminal afferents from the perforant pathway, with cells of origin in the entorhinal cortex. This system constitutes the major cortical afferent input to the limbic system and has been shown to be severely disrupted in AD (Van Hoesen and Hyman 1990; Honer et al. 1992). The dramatic confinement of the Aβ plaques to the OML in PDAPP brains provides additional evidence that strongly supports the postulate that Aβ deposition is associated with deficits in particular neuronal populations or is associated with region-specific factors.

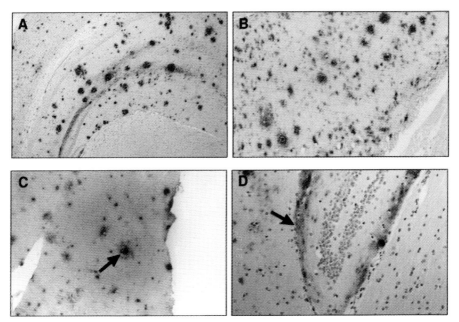

Fig. 1. Aβ plaque pathology in PDAPP mouse brains immunolabeled with human-specific human Aβ antibodies. (A) Numerous diffuse and compacted deposits labeled with the polyclonal Aβ antibody R 1282. Plaques are found throughout the cortical and hippocampal parenchyma of a 14-month-old homozygotic mouse. (B) Aβ cortical plaques immunolabeled with an Aβ $_{X-42}$-specific antibody, 12H7. (C) Double-labeling with Aβ$_{X-42}$-specific (12H7) and Aβ$_{X-40}$-specific (2G3) antibodies reveal that the cores of a small subset of plaques are Aβ$_{X-40}$-positive (arrow). (D) The outer molecular layer of the hippocampal dentate gyrus in a 14-month-old homozygotic mouse is clearly defined by Aβ deposits (arrow).

We can begin to address these issues using the PDAPP mouse model by examining age-associated and regional differences in APP processing and Aβ formation.

APP Processing and Aβ Formation in PDAPP Mouse Brains

Age- and region-dependent amyloid deposition are obvious features of both AD patients and the PDAPP mouse brain. The factors responsible for these observations are presently unknown. Nevertheless, the ability to analyze the PDAPP mouse at various ages, together with the availability of specific antibodies to key intermediates of APP involved in amyloid peptide formation, offer the possibility of determining whether changes in APP expression or metabolism play a role in the amyloid formation and deposition.

The processing pathology of APP is shown in Figure 2. Full-length APP can be cleaved by α-secretase at position 16/17 of the Aβ region giving rise to two fragments: a large N-terminal secreted APP containing the first 16 amino acids at

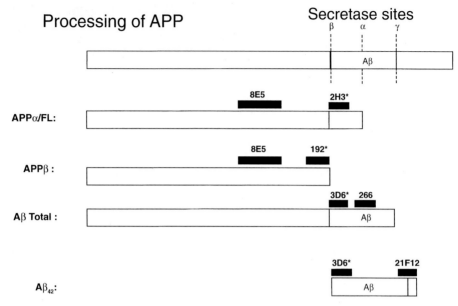

Fig. 2. Immunoassays used to detect APP/Aβ metabolites. Four different immunoassays were used to detect different APP metabolites. In each case sandwich ELISAs employed two different antibodies as indicated where the asterisked antibody was biotinylated and detected with avidin-HRP. Antibodies 192 and 21F12 are affinity-purified rabbit polyclonal antibodies to residues $A\beta^{-1 \text{ to } -5}$ and $A\beta^{34-42}$, whereas all other antibodies indicated are mouse monoclonal antibodies. Antibody 3D6 is specific to the first five amino acids of Aβ and requires a free amino terminus of Aβ whereas 21F12 is specific to the C-terminus of the $A\beta^{42}$ peptides. Each assay, as configured, is highly specific to the indicated APP/Aβ metabolite.

its C-terminus (termed APPα), and a C-terminal fragment with only the latter half of Aβ at its N-terminus (Fig. 2). Clearly this pathway does not lead to Aβ formation and is probably the primary pathway leading to secretion of APP molecules into the conditioned media of cultured cells (Caporasco et al. 1992; Esch et al. 1990) as well as CSF. This pathway has also been shown to be preferentially stimulated by phorbol esters (Oltersdorf et al. 1989).

APP can be processed by a second pathway that involves two rather than one cleavage event (Fig. 2). The first step involves cleavage by the enzyme β-secretase at the Met-Asp site, located at the N-terminus of Aβ. Currently, the identity of this enzyme is not known; however, it is characterized by its exquisite specificity for this cleavage site. Following cleavage at the N-terminus, Aβ is liberated from the resulting C-terminal fragment of APP by a second cleavage event. The C-terminus of Aβ is cut by γ-secretase(s) (Fig. 2). Very little is known about these enzymes; however, the introduction of mutations at position 717 of APP (which corresponds to position 45 of Aβ) results in a shift of the relative processing of $A\beta_{40}$ to $A\beta_{42}$ suggesting that multiple γ-secretases are involved. Importantly, the PDAPP mouse has one such mutation ($717_{V \to F}$).

Over the past several years, a number of sequence-specific monoclonal and polyclonal antibodies to the various APP and Aβ fragments described above have been generated and have greatly facilitated our understanding of APP processing (Seubert et al. 1992, 1993; Fig. 2). We have used these antibodies to generate immunoassays that are specific to unique forms of APP and Aβ to establish whether relative levels of these fragments vary in the PDAPP mouse as a result of age or region.

Based upon the considerations outlined above, four different immunoassays to APP/Aβ were generated (Fig. 2). The first assay uses a binding antibody to the N-terminal domain of Aβ and a reporter antibody to the central of APP. We call this assay APPα/FL, since it detects both APPα as well as full length forms of APP. The second assay uses an antibody ("92") specific to the free C-terminus of APP ending at position 596 of APP ("APPβ") and the same reporter antibody as in the APPα/FL assay. This assay is thus named the APPβ assay. For Aβ, two different assays were also developed. The first assay (Aβ total) sees both $Aβ_{1-40}$ and $Aβ_{1-42}$ and uses a capture antibody directed to the central region of Aβ and a reporter antibody to the amino terminal $Aβ_{1-5}$ epitope. The second immunoassay utilizes a capture antibody (21F12) specific to the C-terminus of $Aβ_{42}$ and therefore recognizes only $Aβ_{1-42}$.

We have used these assays to determine if changes in the levels of Aβ and APP metabolites vary with either the age or brain region of the PDAPP mouse. An analysis of several such measurements is shown in Table 1. The immediate conclusion from the data is that APP levels do not change with the age of the animals whereas Aβ levels are drastically altered in the areas of the brain affected by amyloid. This finding suggests that increased APP expression and processing by β-secretase is unlikely to account for the age-dependent increase in amyloid burden seen in the PDAPP mouse. The second conclusion that can be drawn from the data is that APP levels differ in the different brain regions with hippocampus > cortex > thalamus. The difference in the expression levels is not large, with a maximum change of only two-fold. It is unclear whether this alone can account for the regional deposition of amyloid. Levels of Aβ, either total or 42-specific, strongly correlate with the degree of amyloid deposition from both regional and quantitative perspectives. For example, areas of extensive Aβ amyloidosis, such as the hippocampus and cortex, each reveal marked elevation of Aβ immunoreactivity as measured by the ELISAs. Measurements of APP/Aβ metabolites of the PDAPP mouse collectively suggest that a number of factors might be involved in Aβ amyloidosis. These may include somewhat increased processing in the β-secretase pathway and subtle increases in APP levels in affected brain regions as possible determinants of Aβ deposition.

Neurodegenerative Changes in PDAPP Brains

Extensive neurodegenerative changes occur in the cortical and hippocampal neuropil of AD and PDAPP mouse brains. The abnormal neuritic dystrophy associated with Aβ deposition and the degenerative neuropil in AD has been extensively described (Masliah et al. 1993b; Joachim et al. 1991; Cotman et al. 1991; Arai et al. 1990). During the period of accelerated plaque formation, dystrophic neurites and neuritic plaques develop throughout the affected neuropil in PDAPP mouse brains. Distorted neurites associated with mouse Aβ deposits are immunoreactive with several antibodies recognizing dystrophic abnormalities in AD, including those specific for synaptophysin, MAP 2, GAP 43, heavy and medium chain neurofilaments, and hAPP. Significantly, a subset of the neuritic plaques are associated with neurites and cell bodies immunoreactive with antibodies directed against phosphorylated neurofilaments and tau. They closely resemble human neuritic plaques when viewed by conventional (Fig. 3A), confocal (Fig. 3B) and electron microscopy (Fig. 4). Ultrastructural analysis shows that classic extracellular amyloid fibrillar "stars" and electrodense neuronal processes and cell bodies develop in PDAPP brains, closely resembling those found in AD (Fig. 4). A number of subcellular degenerative changes, including the accumulation of dense laminar and multivesicular bodies, mitochondria and neurofilaments occur in the dystrophic neurites of both PDAPP and AD brains. Interestingly, a number of neurons display nuclei with clumped chromatin and intranuclear inclusions, suggesting that apoptotic-like changes may constitute part of the neurodegenerative changes in PDAPP brains.

The antibody AT 8 recognizes an abnormal phosphorylation site of the microtubule-associated protein tau at Ser-202 that is present in AD brain (Goedert et al. 1993) and has been used extensively to document early cytoskeletal alterations associated with neurofibrillary tangle (NFT) formation in AD (Braak et al. 1994). AT 8 immunoreactivity is thought to precede NFT formation in dystrophic neurites and neuronal cell bodies in AD (Su et al. 1994). In PDAPP brains, AT 8 labels dystrophic neurites in a subset of plaques, thus resembling early tau-associated changes described in AD. However, no paired helical filaments or NFTs have been found in the mouse brains to date. It remains unknown if mice possess the appropriate cellular machinery to generate AD-type NFTs, or whether such changes can occur in rodent brains in less than two years.

Phosphorylated neurofilaments are found in association with tangle-bearing neurons and abnormal neurites (Vickers et al. 1994). The same monoclonal antibodies that label these processes in AD (SMI 312 and SMI 34; Masliah et al. 1993c) label similar structures in PDAPP brains. These include collapsed neuronal cell bodies, atrophied processes and dystrophic neurites that are associated with a subset of hippocampal and cortical Aβ deposits. The accumulation of phosphorylated neurofilaments in PDAPP neuritic processes has also been confirmed by immunogold labeling coupled with electron microscopy.

In addition to the dystrophic changes described above, the loss of immunoreactivity in the hippocampus with antibodies recognizing proteins associated

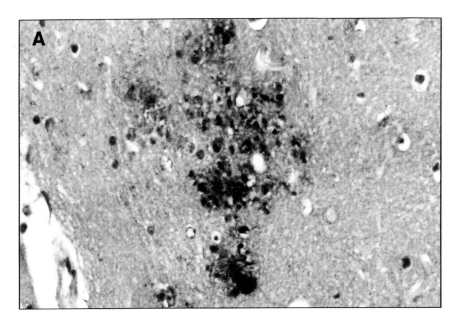

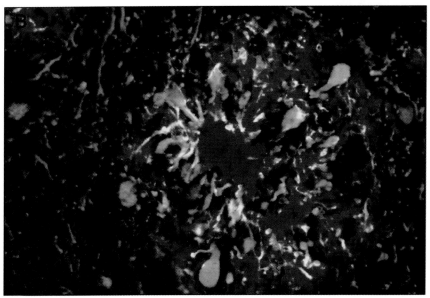

Fig. 3. Neuritic plaque pathology in PDAPP mouse brains. (A) Dystrophic neurites in the hippocampus of a 13-month-old heterozygous mouse labeled with the hAPP-specific antibody 8E5. (B) Neurofilaments (green) associated with neurites in an Aβ deposit (red) in the hippocampus of a 12-month-old heterozygotic mouse.

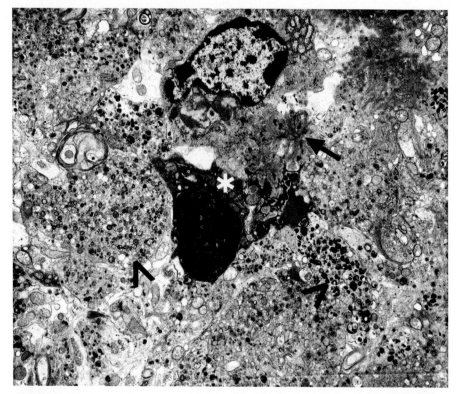

Fig. 4. Electron micrograph of a PDAPP hippocampal neuritic plaque. Extracellular amyloid fibrils (arrowhead) in the vicinity of a degenerating cell body (asterisk) and distended neurites containing multilaminar and electrodense bodies (arrowheads).

with presynaptic vesicles has been documented in PDAPP brains and AD (Honer et al. 1992; Heinonen et al. 1995). Notably, the decrease is pronounced in the outer molecular layer of the hippocampal dentate gyrus in older mice, replicating the pattern of loss in AD (Terry et al. 1994). A decrease in the immunostaining with antibodies recognizing the dendritic marker MAP 2 also occurs in this region, indicating alterations of both presynaptic and postsynaptic elements are present in PDAPP brains.

Gliotic Changes

Both astrocytosis and microgliosis occur in the brains of PDAPP mice. Hypertrophic astrocytes are found associated with the mouse Aβ (Fig. 5A), encircling the plaque in characteristic fashion typical of astrocytosis in AD (Van Eldik and Griffin 1994; Pike et al. 1995; Cairns et al. 1992). Astrocytic gliosis also occurs throughout the affected cortical and hippocampal parenchyma (Fig. 5B, C).

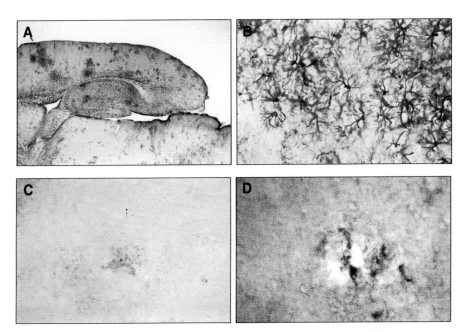

Fig. 5. Gliotic changes in PDAPP mouse brains. (A) Astrocytosis associated with Aβ deposition in an 8-month-old homozygotic mouse. (B) Cortical gliosis in the transgenic brain shown in (A), compared to a non-transgenic littermate (C). (D) Microglial cell cluster immunolabeled with an anti-Mac 1 antibody (Thioflavin S counterstain).

Microglial cells are intimately associated with Aβ deposits in AD and have been identified using a variety of antibodies recognizing histocompatibility complex molecules and other immune system-associated antigens (McGeer et al. 1987; Rogers et al. 1988). Microglial cells in PDAPP mice are labeled with the antibody Mac 1 (Fig. 5D) that recognizes the C3bi complement receptor. The antigen is also expressed in the reactive microglial of AD (Akiyama and McGeer 1990). In PDAPP plaques, microglial cell clusters loosely surround or permeate the core of the deposit, similar to the spatial relationship of microglial cells to the Aβ plaques in AD (Perlmutter et al. 1991).

In summary, several of the major pathological hallmarks associated with AD are present in PDAPP mice overexpressing a mutation associated with early-onset disease. Amyloid plaques in PDAPP mice share several of the major immunoreactive and tinctorial properties definitive of Aβ deposits in AD. Furthermore, massive neurodegenerative changes occur in the brains of these mice that mirror several of the dystrophic and inflammatory lesions associated with AD. These include the loss of immunoreactivity with antibodies against synaptic and dendritic proteins, regional specificity of these lesions, similar subcellular degeneration, abnormal phosphorylation of cytoskeletal elements, reactive astrocytosis and microgliosis and the deposition of lysosomal and acute phase proteins. No neurofibrillary tangles or paired helical filaments have been found in the mice to date. It remains unknown whether mice are capable of generating these in a manner comparable to AD in less than two years. Extensive behavioral analyses are currently being performed in these mice with the intent of correlating cognitive with pathological changes.

Aβ and hAPP can be monitored in these animals, thereby serving as extremely effective tools for testing potential Aβ-reducing therapeutics before their introduction into human trials. Preliminary results suggest that changes in hAPP expression levels with age do not account for the accumulation of Aβ deposits during disease progressing in AD. Regional differences in metabolism, local factors or deficits in distinct populations of neurons are more likely to play a role in the process of amyloidosis. In addition, a number of neurodegenerative and inflammatory changes provide the additional opportunity to test neuroprotective and anti-inflammatory therapeutics.

In conclusion, the PDAPP mouse model provides compelling evidence for the primacy of APP processing and Aβ production in AD neuropathology, and offers novel opportunities to test therapeutics directed towards specific aspects of the neurodegenerative process. In addition to human AD patients, the deposition of Aβ occurs in similar subregions of the brain in other mammalian species. This biological vulnerability is conserved, accelerated and reproducible in PDAPP mouse brains, and probably reflects mechanisms of disease underlying AD pathology.

Acknowledgments

The authors would like to thank T. Guido, K. Khan, M. Mallory and F. Soriano for the preparation of immunohistological material, A. Sisk for the preparation of ultrastructural specimens, R. Barbour, G. Gordon and K. Hu for antibody and assay development and K. Philipkoski for preparation of the manuscript. The antibody R 12802 was generously provided by Dr. Dennis Selkoe.

References

Abraham CR, Selkoe DJ, Potter H (1988) Immunohistochemical identification of the serine protease inhibitor alpha 1-antichymotrypsin inhibitor in the brain amyloid deposits of Alzheimer's disease. Cell 52: 487–501

Akiyama H, McGeer PL (1990) Brain microglia constitutively express β-2 integrins. J Neuroimmunol 30: 81–93

Arai H, Lee VM, Otvos LJr, Greenberg BD, Lowery DE, Sharma SK, Schmidt ML, Trojanowski JQ (1990) Defined neurofilament, tau, and beta-amyloid precursor protein epitopes distinguish Alzheimer from non-Alzheimer plaques. Proc Natl Acad Sci USA 87: 2249–2253

Asian PS, Davis KL (1994) Inflammatory mechanisms in Alzheimer's disease: implications for therapy. Am J Psych 151: 1105–1113

Braak H, Braak E (1995) Staging of Alzheimer's disease-related neurofibrillary changes. Neurobiol Aging 16: 271–278

Braak E, Braak H, Mandelkow EM (1994) A sequence of cytoskeletal changes related to the formation of neurofibrillary tangles and neuropil threads. Acta Neuropath 87: 554–567

Cairns NJ, Chadwick A, Luthert PJ, Lantos PL (1992) Astrocytosis, beta A4-protein and paired helical filament formation in Alzheimer's disease. J Neurol Sci 112: 68–75

Caporasco L, Gandy SE, Buxbaum JD, Ramabhadran TV, Greengard P (1992) Protein phosphorylation regulates secretion of Alzheimer's β/A4 amyloid precursor protein. Proc Natl Acad Sci USA 89, 3055–3059

Cataldo AM, Hamilton DJ, Nixon RA (1994) Lysosomal abnormalities in degenerating neurons link neuronal compromise to senile plaque development in Alzheimer disease. Brain Res 640: 68–80

Cotman CW, Cummings BJ, Whitson JS (1991) The role of misdirected plasticity in plaque biogenesis and Alzheimer's disease pathology. In: Growth factors and Alzheimer's disease. Hefti F, Brachet P, Will B, Christen Y (eds) Alzheimer's disease and related conditions. Springer-Verlag, Heidelberg, pp. 222–233

Cummings BJ, Su JH, Cotman CW, White R, Russell MJ (1993) Beta-amyloid accumulation in aged canine brain: a model of early plaque formation in Alzheimer's disease. Neurobiol Aging 14: 547–560

Esch FS, Keim PS, Beattie EC, Blacher RW, Culwell AR, Oltersdorf T, McClure D, Ward PJ (1990) Cleavage of amyloid β peptide during constitutive processing of its precursor. Science 248: 1122–1128

Fredrickson RC (1992) Astroglia in Alzheimer's disease. Neurobiol Aging 13: 239–253

Fukumoto H, Asami-Odaka A, Suzuki N, Shimada H, Ihara Y, Iwatsubo T (1996) Amyloid β protein in normal aging has the same characteristics as that in Alzheimer's disease. Am J Pathol 148: 259–265

Games D, Adams D, Alessandrini R, Barbour R, Berthelette P, Blackwell C, Carr T, Clemens J, Donaldson T, Gillespie R, Guido T, Hagopian S, Johnson-Wood K, Khan I, Lee M, Leibowitz P, Liebergurb I, Little S, Masliah E, McConlogue L, Montoya Azvala M, Mucke L, Paganini L, Penniman E, Power M, Schenk D, Seubert P, Snyder B, Soriano F, Tan H, Vitale J, Wadsworth S, Wolozin B, Zhao J (1995) Development of neuropathology similar to Alzheimer's disease in transgenic mice overexpressing the $717_{V \rightarrow F}$ β-amyloid precursor protein. Nature 373: 523–527

Goedert M, Jakes R, Crowther RA, Six J, Lubke U, Vandermeeren M, Cras P, Trojanowski JQ, Lee VM (1993) The abnormal phosphorylation of tau protein at Ser-202 in Alzheimer's disease recapitulates phosphorylation during development. Proc Natl Acad Sci USA 90: 5066–5070

Heinonen O, Soininen H, Sorvari H, Kosunen O, Paljarvi L, Koivisto E, Riekkinen PJ (1995) Loss of synaptin-like immunoreactivitiy in the hippocampal formation as an early phenomenon in Alzheimer's disease. Neuroscience 64: 375–384

Higgins LS, Holtzman DM, Rabin J, Mobley WC, Cordell B (1994) Transgenic mouse brain histophathology resembles early Alzheimer's disease. Ann Neurol 35: 598–607

Honer WG, Dickson DW, Gleeson J, Davies P (1992) Regional synaptic pathology in Alzheimer's disease. Neurobiol Aging 13: 375–382

Hyman BT (1992) Down Syndrome and Alzheimer's disease: Progress in clinical and biological research. Prog Clin Biol Res 379: 123–142

Hyman BT, Van Hoesen BW, Damasio AR (1990) Memory-related neural systems in Alzheimer's disease: An anatomical study. Neurology 40: 1721–1730

Iwatsubo T, Mann DMA, Odaka A, Suzuki N, Ihara Y (1995) Amyloid β protein (Aβ) deposition: Aβ 42 (43) precedes Aβ 40 in Down Syndrome. Ann Neurol 37: 294–299

Joachim C, Games D, Morris J, Ward P, Frenkel D, Selkoe D (1991) Antibodies to non-beta regions of the beta-amyloid precursor protein defect a subset of senile plaques. Am J Pathol 138: 373–384

Lemere CA, Blusztajn Y, Yamaguchi T, Wisniewski T, Saido TC, Selkoe DJ (1996) Sequence of deposition of heterogeneous amyloid β-peptides and Apo E in Down syndrome: Implications for initial events in amyloid plaque formation. Neurobiol Disease 3: 16–32

Mann DA (1988) The pathological association between Down Syndrome and Alzheimer's disease. Mech Aging Develop 43: 99–136

Mann DMA (1989) Cerebral amyloidosis, aging and Alzheimer's disease: A contribution of studies from Down's Syndrome. Neurobiol Aging 10: 397–399

Mann DMA (1994) Alzheimer's disease: Progress in pathological and aetiological aspects. Res Gerontol 4: 43–60

McGeer PL, Itagaki S, Tago H, McGeer EG (1987) Reactive microglial in patients with senile dementia of the Alzheimer type are positive for the histocompatibility glycoprotein HLA-DR. Neurosci Lett 79: 195–200

Masliah E, Miller A, Terry RD (1993a) The synaptic organization of the neocortex in Alzheimer's disease. Medical Hypothesis 41: 334–340

Masliah E, Mallory M, Deerink T, DeTeresa R, Lamont S, Miller A, Terry R, Carragher B, Ellisman M (1993b) Re-evaluation of the structural organization of neuritic plaques in Alzheimer's disease. J Neuropathol Exp Neurol 52: 619–632

Masliah E, Mallory M, Hansen L, Alford M, DeTeresa R, Terry R (1993c) An antibody against phosphorylated neurofilaments identifies a subset of damaged association axons in Alzheimer's disease. Am J Pathol 142: 871–882

Murrell J, Farlow M, Ghetti B, Benson M (1991) A mutation in the amyloid precursor protein associated with hereditary Alzheimer's disease. Science 254: 97–99

Oltersdorf T, Fritz LC, Schenk DB, Lieberburg I, Johnson-Wood K, Beattie EC, Ward PJ, Blacher RW, Dovey HF, Sinha S (1989) The secreted form of the Alzheimer's amyloid precursor protein with the Kunitz domain is Protease Nexin II. Nature 341: 144–147

Perlmutter LS, Scott SA, Chui HC (1991) The role of microglia in the cortical neuropathology of Alzheimer disease. Bull Clin Neurosci 56: 120–130

Pike CJ, Cummings BJ, Cotman CW (1995) Early association of reactive astrocytes with senile plaques in Alzheimer's disease. Exper Neurol 132: 172–179

Price DL, Martin LJ, Sisodia SS, Wagner MV, Koo EH, Walker LC, Koliatos VE, Cork LC (1991) Aged non-human primates: An animal model of age-associated neurodegenerative disease. Brain Pathol 1: 287–296

Rockenstein EM, McConlogue L, Tan H, Power M, Masliah E, Mucke L (1995) Levels and alternative splicing of amyloid β protein precursor (APP) transcripts in brains of APP transgenic mice and humans with Alzheimer's disease. J Biol Chem 270: 28257–28267

Rogers J, Luber-Narod J, Styren SD, Civin WH (1988) Expression of immune-system-associated antigens by cells of the human central nervous system: Realationship to the pathology of Alzheimer's disease. Neurobiol Aging 9: 339–349

Selkoe D (1991) The molecular pathology of Alzheimer's disease. Neuron 6: 487–498

Selkoe DJ, Bell DS, Podlinsky MB, Price DL, Cork LC (1987) Conservation of brain amyloid in aged mammals and humans with Alzheimer's disease. Science 235: 873–877

Seubert P, Vigo-Pelfrey C, Esch F, Lee M, Dovey H, Davis D, Sinha S, Schlossmacher M, Whaley J, Swindlehurst C, McCormack R, Wolfert F, Selkow D, Lieberburg I, Schenk D (1992) Isolation and quantification of soluble Alzheimer's β-peptide from biological fluids. Nature 359: 325–327

Seubert P, Oltersdorf T, Lee MG, Barbour R, Blomquist C, Davis DL, Bryant K, Fritz LD, Galasko D, Thal LJ, Lieberburg I, Schenk DB (1993) Secretion of β-amyloid precursor protein cleaved at the amino terminus of the β-amyloid peptide. Nature 361: 260–263

Snow AD, Mar H, Nochlin D, Kimata K, Kato M, Suzuki S, Hassell J, Wight TN (1988) The presence of heparin sulfate proteoglycans in the neuritic plaques and congophilic angiopathy of Alzheimer's disease. Am J Pathol 133: 456–463

Su JH, Cummings BJ, Cotman CW (1994) Early phosphorylation of tau in Alzheimer's disease occurs at Ser-202 and is preferentially located within neurites. NeuroReport 5: 2358–2362

Takahashi H, Kurashima C, Utuyama M, Hirokawa K (1990) Immunohistological study of senile brains using a monoclonal antibody recognizing beta amyloid precursor protein, significance of granular deposits in relation with senile plaques. Acta Neuropathol 80: 260–265

Terry RD, Masliah E, Hansen LA (1994) Structural basis of the cognitive alterations in Alzheimer's disease. In: Terry RD, Katzman R, Bick KL (eds). Alzheimer's disease. Raven Press, New York pp. 179–196

Van Eldik LJ, Griffin WS (1994) S100 beta expression in Alzheimer's disease: relation to neuropathology in brain regions. Biochim Biophys Acta 1223: 398–403

Van Hoesen GW, Hyman BT (1990) Hippocampal formation: Anatomy and the patterns of pathology in Alzheimer's disease. Prog Brain Res 83: 445–457

Vickers JC, Riederer BM, Marugg RA, Buee-Scherrer V, Buee L, Delacourte A, Morrison JH (1994) Alterations in neurofilament protein immunoreactivity in human hippocampal neurons related to normal aging and Alzheimer's disease. Neuroscience 62: 1–13

Wisniewski HM, Bancher C, Barcikowska M, Wen GY, Currie J (1989) Spectrum of morphological appearance of amyloid deposits in Alzheimer's disease. Acta Neurophathol 78: 337–347

Yamaguchi H, Harai S, Morimatsu M, Shoji M, Ihara Y (1988) A variety of cerebral amyloid deposits in the brains of Alzheimer-type dementia demonstrated by β protein immunostaining. Acta Neuropathol 541–549

Molecular Mechanisms of Synaptic Disconnection in Alzheimer's Disease

E. Masliah[*], M. Mallory, M. Alford, R. DeTeresa, A. Iwai, and T. Saitoh

Abstract

Synaptic loss and neurofibrillary pathology are major contributors to the cognitive deficits in Alzheimer's disease (AD), indicating an altered connectivity of association neurocircuitries. Synaptic damage occurs early in the development of AD, suggesting that synapse pathology is a primary rather than a secondary event. The mechanisms of synaptic damage and neurodegeneration in AD are not completely understood. Recent studies have suggested that abnormal expression and/or processing of growth-associated proteins in the central nervous system might play a role in the mechanisms leading to synaptic damage and neurodegeneration in AD. Prominent among these proteins are amyloid precursor protein (APP), apolipoprotein E (apoE), and non Aβ amyloid component (NAC) precursor (NACP). All of these molecules have several common features: 1) modulation of synaptic function, 2) involvement in amyloidogenesis, and 3) mutations (APP) and polymorphisms (APOE, NACP) that are associated with a higher risk for AD. Abnormal functioning of synaptic-related proteins with amyloidogenic potential might play a central role in the pathogenesis of AD. In this context, the main objectives of this manuscript are to review the contribution of synaptic alterations to the mechanisms of dementia in AD and to discuss some of the possible mechanisms through which malfunctioning of APP, apoE and NACP might lead to synaptic damage and plaque formation in AD.

Introduction

The cognitive alterations in Alzheimer's disease (AD) are associated with widespread neurodegeneration throughout the association cortex and limbic system (Braak and Braak 1991; Heinonen et al. 1995; Hof and Morrison 1994; Hyman et al. 1986; Masliah and Terry 1994; Masliah 1995a; Masliah et al. 1993a). This neurodegenerative process is characterized by synaptic and neuronal loss (DeKosky and Scheff 1990; Masliah et al. 1994; Terry et al. 1981, 1991), plaque

---

[*] Departments of Neurosciences and Pathology, University of California, San Diego, School of Medicine, La Jolla, California 92093–0624

B. T. Hyman / C. Duyckaerts / Y. Christen (Eds.)
Connections, Cognition, and Alzheimer's Disease
© Springer-Verlag Berlin Heidelberg 1997

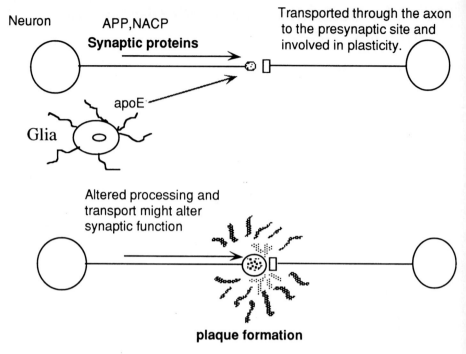

Fig. 1. Role of synaptic proteins in Alzheimer's disease pathophysiology.

(Yamaguchi et al. 1988) and tangle formation (Arriagada et al. 1992a), concomitant decrease in specific neurotransmitters (Cowburn et al. 1990; Perry et al. 1977) and gliosis (Beach et al. 1989). Synaptic loss (DeKosky and Scheff 1990; Terry et al. 1991) and neurofibrillary pathology (Arriagada et al. 1992a) are major contributors to the cognitive deficits in AD, indicating an altered connectivity of association neurocircuitries (Hof and Morrison 1994; Masliah et al. 1993a, b; Samuel et al. 1994a, b). Synaptic damage occurs early in the development of AD (Heinonen et al. 1995; Masliah et al. 1994), suggesting that synapse pathology is a primary rather than a secondary event, yet the mechanisms of synaptic damage and neurodegeneration in AD are not completely understood. Recent studies have suggested that abnormal expression and/or processing of growth-associated proteins in the central nervous system (CNS) may play a role in the mechanisms leading to synaptic damage and neurodegeneration in AD (Masliah et al. 1991b; Masliah and Terry 1993; Masliah 1995a; Fig. 1). Prominent among these are amyloid precursor protein (APP; Masters et al. 1985; Selkoe 1989, 1993), apolipoprotein E (apoE; Roses et al. 1996), and non-Aβ amyloid component (NAC) precursor (NACP; Masliah et al. 1995a). All of these molecules have several common features (Table 1): 1) the ability to modulate synaptic function, 2) involvement in amyloidogenesis, and 3) mutations (e.g.; APP) and polymorphisms (e.g.; APOE, NACP) that are associated with a higher risk for AD

Table 1. Role of synapse-associated proteins in Alzheimer's disease

Protein	Cromosomal location	Synaptic localization	Plaque localization
APP	21	3+	5+
apoE	19	2+ (after injury)	3+
NACP	4	5+	3+
Presenilins	1, 14	dendritic	3+

(Table 1). More recently, a group of new proteins called presenilins (PS 1, PS 2; Collaborative 1995; Kovacs et al. 1996) has been implicated in the pathogenesis of AD; however, the role of these molecules in synaptic function is currently unknown (Table 1).

The main objectives of this manuscript are to review the contributions of synaptic alterations to the mechanisms of dementia in AD and to discuss some of the possible mechanisms through which malfunctioning of APP, apoE and NACP might lead to synaptic damage and plaque formation in AD.

The Role of Synaptic Alterations in Mechanisms of Dementia in AD

Recent studies shown that, in addition to the traditionally described lesions (plaques and tangles) found in the AD brain (Alzheimer 1907; Braak and Braak 1991; Dickson et al. 1988; Terry et al. 1964; Terry and Wisniewski 1970; Yamaguchi et al. 1988), this neurodegenerative disease is characterized by neuronal loss (Hof et al. 1990; Terry et al. 1981), disruption of the neuritic cytoskeleton with altered cortico-cortical connectivity (Hof et al. 1990; Masliah et al. 1993a; Morrison et al. 1987), and extensive synapse loss (Davies et al. 1987; DeKosky and Scheff 1990; Hamos et al. 1989; Honer et al. 1992; Lassmann et al. 1992; Masliah et al. 1989, 1991a, c). It has been hypothesized that in AD the dementia could be caused by either the individual presence of these specific lesions or by the synergistic effect of some or all of these lesions (DeKosky and Scheff 1990; Samuel et al. 1994a, b; Terry et al. 1991; Fig. 2). The original studies by Blessed and Tomlinson suggested that amyloid deposition and plaque formation might be the major correlates of cognitive alterations in AD (Blessed et al. 1968), but more detailed studies where control cases were not included in the linear regression analysis did not support this view (DeKosky and Scheff 1990; Dickson et al. 1995; Terry et al. 1991). Other groups have shown that neuronal loss in specific areas of the neocortex and subcortical regions correlated with clinical alterations seen in AD (Neary et al. 1986). However, these correlations are rather weak and do not completely explain all the clinical alterations observed in AD. An alternative hypothesis is that dementia in AD is directly associated with the disruption of neuritic substructure and loss of synaptic connection in specific neocortical, limbic and subcortical areas (Masliah et al. 1991c; McKee et al. 1991). Measure-

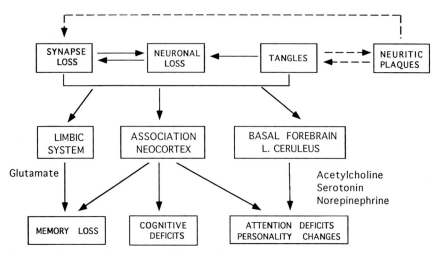

Fig. 2. Mechanisms of dementia in Alzheimer's disease.

ments at both electron microscopic and immunocytochemical levels have shown strong correlations between synaptic numbers in the frontal cortex and test scores of global cognition in AD (DeKosky and Scheff 1990; Terry et al. 1991). More recently, correlative studies between tests scores of cognition and immunochemical quantification of various synaptic proteins have confirmed this view (Dickson et al. 1995; Lassmann et al. 1992; Zhan et al. 1993). In the neocortex, synaptic damage results in retrograde degeneration of cortico-cortical connections between association neurons in layers 2 and 3 (Hof et al 1990; Hof and Morrison 1994; Masliah et al. 1993a, b; Fig. 3). This circuit is believed to be involved in higher cognitive functions (Eccles 1981, 1984). Moreover, the system of horizontal fibers that connect the neurons in layer 6 with those in layer 1 displays significant damage (Masliah et al. 1993a; Fig. 3). This circuit is believed to be involved in memory (Eccles 1981, 1984).

Neurodegeneration of association axons in the neocortex might result in deficits in learning, language, calculation, naming and orientation (Terry et al. 1991; Fig. 2), whereas synaptic damage and axonal injury in the limbic system and basal forebrain leads to memory deficits (Hyman et al. 1986; Fig. 2). Limbic circuitries affected in AD include the perforant pathway (Hyman et al. 1987) and CA-dentate granular cell connections with CA3, CA3 with CA1 and CA1 with subiculum (Samuel et al. 1994a; Fig. 4). Synapse loss was most strongly correlated with dementia when it occurred in the molecular layers of the dentate fasciculus and stratum lacunosum, CA2/3, and CA4; synapse loss in these subregions appeared significantly clustered on factor analysis. In general, these results were compatible with a two-component model of hippocampal connectivity and function in the context of AD. The first component consists of subregions preceding CA1 in a hypothesized input-processing sequence intrinsic to the hippocampus that summates neuronal excitation and influences cognition primarily through

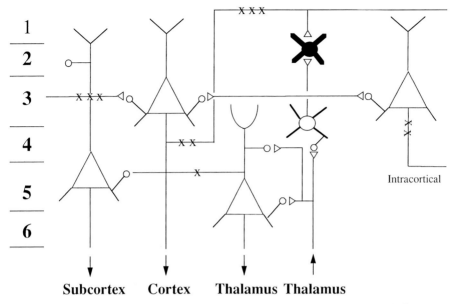

Fig. 3. Neocortical neuronal circuitries interrupted in Alzheimer's disease. Cortico-cortical connections within and among modules are affected. The number of "x" indicates the severity of injury.

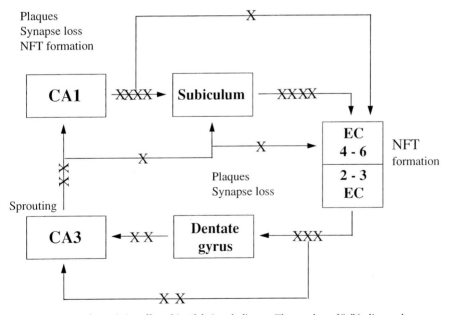

Fig. 4. Hippocampal circuitries affected in Alzheimer's disease. The number of "x" indicates the severity of injury.

synapse density. The second component consists of an "output module," mainly CA1 and the subiculum, that receives the processed signal and passes it on to extrahippocampal cortical and subcortical targets, and affects cognition primarily by NFT accumulation in output neurons. A "net pathology" score combining standardized z-scores for synapse density and NFTs was significantly correlated with all three mental status measures in all hippocampal subregions except the entorhinal cortex, and stepwise regression on these data found net pathology in CA4 to be the most independent significant predictor of dementia (Samuel et al. 1994a).

Synaptic pathology in AD could be either the direct (or primary) result of an underlying molecular defect affecting the synapses, or an indirect (or secondary) result of neuronal loss, plaque and tangle formation (Masliah and Terry 1994; Masliah 1995a). In this regard it is worth noting that neurofibrillary pathology in the neocortex, as assessed by TG3 immunoreactivity, and or/by quantification of neuritic plaques (McKee et al. 1991), neuropil threads (Delaere et al. 1989; Masliah et al. 1992a; Samuel et al. 1994b) or NFTs (Arriagada et al. 1992a b) is an important correlate to the cognitive alterations. Stepwise regression analysis suggests that synaptic and neurofibrillary alterations contribute up to 70 % of the strength of the correlations, whereas amyloid deposits contribute up to the 10 % of the correlation with the cognitive performance expressed as Blessed score (Masliah 1995a). To further clarify the contribution of each of these factors to the dementia in AD, recent studies have focused on characterizing cases that reflect the progression of the disease (Masliah 1995b). These studies have shown that in very early AD cases there is a mild synaptic loss in the outer molecular layer of the dentate gyrus which is accompanied by abnormal tau and APP immunoreactivity in the entorhinal cortex, as well as diffuse amyloid deposition in this region (Braak and Braak 1991; Heinonen et al. 1995; Masliah et al. 1994; Masliah 1995b; Morris et al. 1996; Table 2). Furthermore, there is no neuronal loss in the limbic system or neocortex in these cases and only mild astrocytic and microglial reaction is observed (Table 2). In the later stages of the disease there is NFT formation, first in the limbic system and then in the neocortex, accompanied by formation of neuritic plaques and accentuation of synaptic pathology (Braak and Braak 1991; Masliah et al. 1994). Taken together these results suggest that altered functioning of vulnerable neurons might result in synaptic loss which in turn is associated with cognitive dysfunction and amyloid deposition (Fig. 5). Although

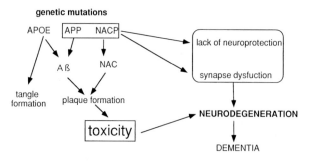

Fig. 5. Mechanisms of neurodegeneration in Alzheimer's disease.

Table 2. Summary of progression of the neuropathological alterations in Alzheimer's disease.[a]

	Incipient	Mild	Moderate	Severe
Neocortex				
Neuronal loss	–	–	+	++
Synapse loss	–	+	++	+++
Pretangles	–	–	+	+
Tangles	–	–	+	++
Diffuse plaques	+/–	+	+++	++
Neuritic plaques	–	+/–	+	++
Limbic system				
Neuronal loss	–	–	+	++
Synapse loss	+	++	+++	++++
Pretangles	+	++	++	++
Tangles	+/–	+	++	+++
Diffuse plaques	++	++++	+++	+++
Neuritic plaques	–	+/–	++	+++

[a] – = none; 1+ = rare; 2+ = occasional; 3+ = abundant; 4+ = very abundant

β-amyloid (Aβ) deposition may or may not be the initial triggering factor, the eventual accumulation of the deposits throughout the brain may contribute to the overall degenerative process.

Possible mechanisms involved in the pathogenesis of synaptic damage in AD could be related to either abnormal function of synaptic proteins or direct toxic effects at the presynaptic site (Masliah and Terry 1993; Masliah 1995a; Fig. 5). Alternatively, abnormal functioning of cytoskeletal proteins might result in abnormal axonal flow, which in turn might lead to synaptic damage (Suzuki and Terry 1967). Recent studies have shown that APP, which is believed to be centrally involved in AD (Selkoe 1989), might play an important role as a synaptic regulator (Alvarez et al. 1992; Askanas et al. 1992; Roch et al. 1994; Schubert et al. 1991; Small et al. 1994). Moreover, APP metabolism appears to be abnormal in AD (Sisodia et al. 1990; Zhong et al. 1994). Furthermore, recent studies in transgenic mice have shown that overexpression of mutated human APP is accompanied by synaptic damage, neurodegeneration and plaque formation (Games et al. 1995). Taken together these findings suggest that altered APP processing may lead to synaptic dysfunction in AD.

The Role of APP in the Pathogenesis of Synaptic Damage in AD

Amyloid precursor protein is a 110 to 120 kDa glycoprotein encoded by chromosome 21 that is centrally involved in AD pathogenesis (Masters et al. 1985; Selkoe 1989) because mutations within this molecule are associated with familial AD (Clark and Goate 1993; Goate et al. 1991) and overexpression of mutated APP in transgenic mice results in AD-like pathology (Games et al 1995). To better understand the role of APP in AD pathophysiology, it is important to understand the function of this molecule in the normal CNS. In this regard, recent studies have shown that APP is found primarily in neurons (Arai et al. 1991; Card et al. 1988; Shimokawa et al. 1993) with a preferential localization at central and peripheral synaptic sites (Askanas et al. 1992; Masliah et al. 1992c; Schubert et al. 1991), suggesting a possible role in neuroplasticity (Mucke et al. 1994; Saitoh et al. 1994). Furthermore, studies have shown that secreted APP (sAPP) fulfills synaptotrophic (Masliah et al. 1995b; Mucke et al. 1994; Roch et al. 1994) and neuroprotective functions within the CNS in response to excitotoxicity (Masliah et al. 1997a; Mucke et al. 1995; Mattson et al. 1993a) and ischemia (Bowes et al. 1994; Mattson et al. 1993a; Smith-Swintosky et al. 1994). Therefore, abnormal functioning of sAPP may be involved in the mechanisms of synaptic damage by failing to promote or maintain normal synaptic populations after excitotoxic challenge (Masliah et al. 1994; Masliah 1995a; Mattson et al. 1993a).

Secreted forms of APP are generated by proteolytic cleavages within their extracellular domain close to the transmembrane region (Kounnas et al. 1995; Tanzi et al. 1988). The extracellular regions of APP751, 770 and APP-like protein 2 (APLP2) each contain a Kunitz protease inhibitor (KPI; Tanzi et al. 1988). Secreted forms of APP containing the KPI are also known as protease nexin II (PNII; Van Nostrand et al. 1989). Secreted APP binds with high affinity to both fibroblasts and neuronal cells. Recent studies have shown that the KPI-containing form of sAPP binds to the low density lipoprotein (LDL) receptor-related protein (LRP; Kounnas et al. 1995). LRP is a multiligand receptor and a member of the LDL receptor family (Strickland et al. 1995). In the CNS, LRP also binds apoE (Strickland et al. 1995) and is present in neurons, microglia and reactive (but not resting) astrocytes (Rebeck et al. 1993). APP, APLP and LRP are colocalized and concentrated at the synaptic site, which suggests a role in balancing synaptic integrity (Kounnas et al. 1995). Disruption of this mechanism could result in synaptic damage in AD (Masliah 1995a).

Abnormal APP functioning could lead to neurodegeneration in AD by 1) direct toxicity of elevated levels of aggregated Aβ (Golde et al. 1992; Selkoe 1993), 2) disruption of synaptic function (Masliah 1995a), 3) deficient neuroprotective activity against excitotoxicity (Mattson et al. 1993a; Mucke et al. 1995), and 4) any combination of the three (Fig. 5). The direct toxic role of fibrillar Aβ in *in vivo* is controversial. In addition, the amyloid load in the brain of patients with AD, in general, does not correlate with the cognitive deficits and with the extent of the neurodegenerative process (Arriagada et al. 1992a, b; DeKosky and Scheff 1990; Terry et al. 1991), with the exception of a recent study by Cummings and Cotman (1995). Therefore, although Aβ might play a limited role in promoting neuronal dysfunction, an additional process such as excitotoxicity might be required to satisfactorily explain AD pathology. It has been postulated that sAPP might prevent the toxicity associated with activation of glutamate receptors by stabilizing intracellular calcium levels (Mattson et al. 1993a, b, c). Furthermore, neuroprotection might also be achieved by modulating the levels of glutamate receptors and/or activity of glutamate-aspartate uptake systems. With respect to the latter, studies have shown that potentially neurotoxic neurotransmitters like glutamate are cleared from the synaptic cleft by high affinity, Na^+-dependent uptake carriers located in both neurons and glia (Balcar and Li 1992; Greenamyre and Porter 1994; Kanai et al. 1993, 1994; Fig. 6). The high affinity uptake system for glutamate and aspartate is heterogeneous (Balcar and Li 1992). There are at least two classes of high affinity uptake systems: 1) a Na^+dependent, Cl-independent system, which takes up both glutamate and aspartate and 2) a Na^+-independent, Cl-dependent system, which takes up glutamate exclusively. This has been determined based on binding studies utilizing radiolabeled D- and L-aspartate and glutamate (Balcar and Li 1992; Cross et al. 1986). Recently, at least three high affinity Na^+-dependent uptake carriers of aspartate/glutamate [also known as glutamate transporters (GTs)] have been cloned: GLT-1 (or EAAT2; Pines et al. 1992), EAAC1 (Kanai and Hediger 1992), GLAST (Storck et al. 1992; or EAAT1) and the cerebellar transporter (or EAAT4; Kanai et al. 1995). These transporters exhibit 39–55% sequence homology with each other (Kanai et al. 1995). GLT-1 is specifically located in astrocytes, EAAC1 has a neuronal localization that includes nonglutaminergic neurons and GLAST is localized in subsets of neurons and glial cells (Rothstein et al. 1994). Studies of GTs have shown that transport is electrogenic and coupled to the cotransport of two Na^+ ions and the counter-transport of one K^+ and one OH^+ ion (Kanai et al. 1995). It now seems clear that these as well as other not fully characterized GTs are present in specific cellular populations and regions of the brain and are necessary for clearing excitotoxic neurotransmitters (Kanai et al. 1995; Manfras et al., 1994; Shashidharan et al. 1994). Thus, deficient GT activity might result in neurodegeneration by causing accumulation of excitotoxins at the synaptic site (Fig. 6). Supporting a role for decreased GT activity in neurodegeneration, recent studies have shown that: 1) in AD the high affinity glutamate/aspartate uptake system is 40–50% decreased in the neocortex (Cowburn et al. 1988, 1990; Masliah et al. 1996b; Scott et al. 1995), 2) in amyotrophic lateral sclerosis there is selective loss of the glial GT (GLT-1; Rothstein et al. 1995), and 3) chronic inhibition of GT activity with L(-)-threo-3-hydroxy aspartate (THA) in experimental models results in neurodegeneration (Rothstein et al. 1993). In addition, we have recently shown that GT deficits occur early in the progression of AD and are correlated with the extent of the neurodegenerative process as assessed by levels of brain spectrin degradation products (Masliah et al., 1996b). These results suggest that abnormal functioning of APP might be associated with decreased activity of GTs in AD, which in turn leads to increased excitotoxicity and neurodegeneration (Fig. 6).

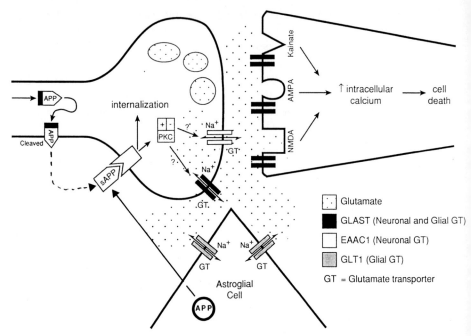

Fig. 6. The role of APP and glutamate transporters (GT) in regulation of glutamate excitotoxicity.

The Role of apoE in the Pathogenesis of Synaptic Damage in AD

Apolipoprotein E is a 34-kDa protein that plays an important role in cholesterol transport, uptake and redistribution (Mahley and Innerarity 1983; Paik et al. 1985). ApoE is encoded by a polymorphic gene located in chromosome 19 (Lin-Lee et al. 1985) and the three different isoforms are encoded by three separate alleles (ε2, ε3 and ε4) that are inherited in a co-dominant fashion at a single genetic locus (Zannis and Breslow 1981). Recent studies showed that approximately 64 % of AD cases are associated with the presence of allele ε4 of APOE (Saunders et al. 1993), suggesting that an abnormally functioning apoE might lead to neurodegeneration in AD (for review see Roses et al. 1996). The precise mechanisms by which abnormal functioning of apoE might lead to AD are not yet fully understood. While some studies support a role of apoE in amyloid clearance (Schmechel et al. 1993; Strittmatter et al. 1993), others suggest a role in protection against excessive tau phosphorylation (Strittmatter et al. 1994), intracellular calcium influx (Hartmann et al. 1994) and synaptic plasticity (Poirier, 1994; Fig. 7).

Supporting a role for apoE in synaptic plasticity, recent *in vitro* studies have shown that apoE complexed with other lipids promotes neuritic outgrowth (Holtzman et al. 1995; Ignatius et al. 1987; Nathan et al. 1994). *In vivo*, apoE mediates delivery of cholesterol to synaptic membranes during the process of

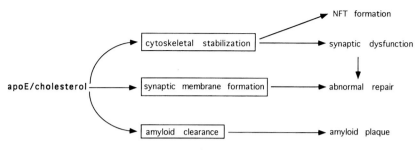

Fig. 7. The role of apolipoprotein E in Alzheimer's disease.

synaptic plasticity and regeneration after injury (Masliah et al. 1995b; Poirier et al. 1993). Furthermore, studies in apoE-deficient mice have shown that this molecule plays an important role in the maintenance of the synapto-dendritic complex during aging (Masliah et al. 1995c, d; 1996a). The neurodegenerative process in the neocortex and limbic system of aged apoE-deficient mice is accompanied by behavioral alterations (Gordon et al. 1995) that are reversed by infusion of apoE (Masliah et al. 1996b). In addition, preliminary studies in a subset of AD cases matched for disease duration have shown that ε4 cases showed a tendency toward greater synaptic damage (Masliah et al. 1996b), compared to cases with ε3 allele. Cholinergic deficits are also more extensive in AD with APOEε4 (Poirier et al. 1995). Analysis of apoE immunoreactivity by Western blot analysis of brain homogenates showed an average 20% decrease in both AD and the Lewy body variant of AD (LBV) cases when compared to controls that correlated with the extent of the synaptic damage (Masliah et al. 1996a). This finding is consistent with previous studies showing a decrease in apoE in the cortex and hippocampus in AD patients with APOEε4 (Bertrand et al. 1995). ApoE mRNA was more abundant in AD compared to controls, but only in APOEε3 cases (Yamada et al. 1995). We conclude that since apoE appears to play an important role in development and regeneration of the nervous system, the expression of the APOEε4 allele might be associated with a defective mechanism of sprouting that eventually accentuates the synaptic alterations in AD.

Supporting this contention, previous studies in dorsal root ganglion cells and in an immortalized CNS-derived neuronal cell line have shown that lipid-complexed human apoE3 increases neuritic outgrowth, whereas apoE4 decreases neuritic outgrowth (Holtzman et al. 1995; Nathan et al. 1994). It has been suggested that the differential effects of apoE isoforms might be related to the interactions of apoE with its receptor at multiple subcellular compartments (Holtzman et al. 1995). Furthermore, kinetic studies with apoE have shown that E3 is metabolized differently from E4, supporting the possibility that the E4 isoform might represent a dysfunctional molecule (Gregg et al. 1986). This is of particular interest for the understanding of the pathogenesis of AD, because a higher risk for this neurodegenerative disorder is associated with the presence of the allele ε4 of APOE (Saunders et al. 1993).

In summary, the results of the various studies reviewed indicate that apoE4 is kinetically different from apoE3, and suggest that the presence of a malfunctioning apoE might play a role in synaptic dysfunction and neurodegeneration (Fig. 7).

The Role of NAC/NACP in the Pathogenesis of Presynaptic Damage in AD

NACP, also called synuclein alpha, is a 19 kDa (Iwai et al. 1994; Ueda et al. 1993) presynaptic protein loosely associated with synaptic vesicles (Iwai et al. 1994) which is encoded by chromosome 4 (Campion et al. 1995; Chen et al. 1995). NACP sequence shows 95 % homology with human synuclein β and with bovine phosphoneuroprotein 14 (Maroteaux and Scheller 1991; Nakajo et al. 1993; Ueda et al. 1993). NACP mRNA is expressed mainly in the brain and to a lower extent in the heart, muscle pancreas and placenta (Ueda et al. 1994). NAC is a 35 amino acid peptide derived from the larger precursor protein, NACP (Ueda et al. 1993). Secondary structure analysis has shown that NAC has a strong tendency to form β-pleated structures (Iwai et al. 1995) and immunocytochemical analysis has shown that antibodies raised against synthetic peptides corresponding NAC sequence recognize amyloid fibrils in the plaques (Masliah et al. 1995a; Ueda et al. 1993; Fig. 8). Furthermore, NAC immunoreactivity is associated mainly with plaques observed in AD, but not with those seen in normal aging (Masliah

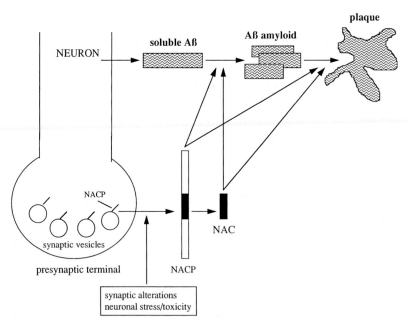

Fig. 8. The role of NACP in synaptic damage and plaque formation in Alzheimer's disease.

et al. 1995a). While NACP is associated with the dystrophic neurites in plaques, NAC is associated with the dense amyloid deposits in the plaque core (Masliah et al. 1995a). Early AD cases showed one-fold higher levels of NACP immuno-reactivity compared to moderate and severe AD (Iwai et al. 1996), and these levels correlated with tangle counts and Blessed score, but not with plaque counts. this study suggests that the abnormal accumulation of NACP during the early stages of AD might play an important role in the mechanisms of neurode-generation and synaptic damage in AD. Genetic association and linkage studies using a polymorphic dinucleotide repeat sequence in the NACP gene showed that one of the NACP polymorphisms (NACP allele 2) has significant association with healthy elderly individuals with APOE risk, indicating a possible protective func-tion of the allele (Xia et al. 1996). Nevertheless, screening of AD families failed to establish linkage between NACP and AD (Brookes et al. 1994; Campion et al. 1995).

Recent studies have suggested that abnormal functioning of NACP might be involved in the pathogenesis of AD by promoting amyloid deposition and inter-fering with synaptic function (Iwai et al. 1994; Masliah et al. 1995a; Fig. 8). As to the first possibility, the high propensity of NAC for amyloid fibrils suggests that the codeposition of NAC and Aβ may represent an early event in plaque forma-tion. In this regard, NAC might differ from other plaque constituents such as ace-tylcholinesterase, immunoglobulin, and heparan sulfate proteoglycan, whose inclusion might be a secondary phenomenon (Brookes et al. 1994). Furthermore, NACP binds Aβ and promotes the aggregation and fibrillar formation (Yoshi-moto et al. 1995). This binding ability is specifically dependent on the 35 amino acid NAC sequence within NACP (Yoshimoto et al. 1995). Transglutaminase cata-lyzes the formation of the covalent NAC polymers, as well as aggregates with Aβ (Jensen et al. 1995). The transglutaminase reactive amino acid residues in NAC were identified as Gln79 and Lys80, the latter localized in a consensus motif con-served in the NACP/synuclein gene family (Jensen et al. 1995). Therefore, trans-glutaminase might participate in amyloid formation and NACP/synuclein modi-fication.

Concluding Remarks

Analysis of molecules associated with amyloid in the core of senile plaques has revealed that many of them including APP, apoE and NACP, are important func-tional molecules of presynaptic terminals. Abnormal functioning of these synapse-related proteins with amyloidogenic potential might play a central role in the pathogenesis of AD (Fig. 5). These molecules might hold future promise for development of novel treatment.

Acknowledgments

This work was supported by NIH Grants AG05131 and AG10689, and with funding by the Alzheimer Disease and Related Disorders Association and the Ruth K. Broad Biomedical Research Foundation.

References

Alvarez J, Moreno RD, Llanos O, Inestrosa NC, Brandan E, Colby T Esch FS (1992) Axonal sprouting induced in the sciatic nerve by the amyloid precursor protein (APP) and other antiproteases. Neurosci Lett 144: 130–134

Alzheimer A (1907) Über eine eigenartige Erkrankung der Hirnrinde. Allgemeine Zeitschrift für Psychiatrie 64: 146–148

Arai H, Lee VM-Y, Messinger ML, Greenberg BD, Lowery DE, Trojanowski JQ (1991) Expression patterns of β-amyloid precursor protein (β-APP) in neural and nonneural tissues from Alzheimer's disease and control subjects. Ann Neurol 30: 686–693

Arriagada PV, Growdon JH, Hedley-Whyte ET, Hyman BT (1992a) Neurofibrillary tangles but not senile plaques parallel duration and severity of Alzheimer's disease. Neurology 42: 631–639

Arriagada PV, Marzloff K, Hyman BT (1992b) Distribution of Alzheimer-type pathologic changes in nondemented elderly individuals matches the pattern in Alzheimer's disease. Neurology 42: 1681–1688

Askanas V, Engel WK, Alvarez RB (1992) Strong immunoreactivity of β-amyloid precursor protein, including the β-amyloid protein sequence, at human neuromuscular junctions. Neurosci Lett 143: 96–100

Balcar VJ, Li Y (1992) Heterogeneity of high affinity of L-glutamate and L-aspartate in the mammalian central nervous system. Life Sci 51: 1467–1478

Beach TG, Walker R, McGeer EG (1989) Patterns of gliosis in Alzheimer's disease and aging cerebrum. Glia 2: 420–436

Bertrand P, Poirier J, Oda T, Finch CE, Pasinetti CM (1995) Association of apolipoprotein E genotype with brain levels of apolipoprotein E and apolipoprotein J (clusterin) in Alzheimer's disease. Mol Brain Res 33: 174–178

Blessed G, Tomlinson BE, Roth M (1968) The association between quantitative measures of dementia and senile change in the cerebral grey matter of elderly subjects. Br J Psych 114: 797–811

Bowes MP, Masliah E, Chen B-L, Otero D, Zivin J, Saitoh T (1994) Reduction of neurological damage by a trophic peptide segment of the amyloid β/A4 protein (APP). Exp Neurol 129: 1–8

Braak H, Braak E (1991) Neuropathological stageing of Alzheimer-related changes. Acta Neuropathol 82: 239–259

Brookes AJ, St. Clair D (1994) Synuclein proteins and Alzheimer's disease. Trends Neurosci 17: 404–405

Campion D, Martin C, Heilig R, Charbonnier F, Moreau V, Flaman JM, Petit JL, Hannequin D, Brice A, Frebourg T (1995) the NACP/synuclein gene: chromosomal assignment and screening for alterations in Alzheimer disease. Genomics 26: 254–257

Card JP, Meade RP, Davis LG (1988) Immunocytochemical localization of the precursor of the precursor protein for β-amyloid in the rat central nervous system. Neuron 1: 835–846

Chen X, de Silva RHA, Pettenati MJ, Rao PN, St.George-Hyslop P, Roses AD, Xia Y, Horsburgh K, Ueda K, Saitoh T (1995) The human NACP/alpha-synuclein gene: chromosome assignment to 4q21.2-q22 and TaqI RFLP analysis. Genomics 26: 425–427

Clark RF, Goate AM (1993) Molecular genetics of Alzheimer's disease. Arch Neurol 50: 1164–1172

Collaborative G (1995) The structure of the presenilin 1 (S182) gene and identification of six novel mutations in early onset AD families. Alzheimer's Disease Collaborative Group. Nature Gen 11: 219–222

Cowburn R, Hardy J, Roberts P, Briggs R (1988) Presynaptic and postsynaptic glutamatergic function in Alzheimer's disease. Neurosci Lett 86: 109–113

Cowburn RF, Hardy JA, Roberts PJ (1990) Glutamatergic neurotransmission in Alzheimer's disease. Biochem Soc Trans 18: 390–392

Cross AJ, Skan WJ, Slater P (1986) The association of [^{3}H]d-aspartate binding and high-affinity glutamate uptake in the human brain. Neurosci Lett 63: 121–124

Cummings BJ, Cotman CW (1995) Image analysis of beta-amyloid load in Alzheimer's disease and relation to dementia severity. Lancet 346: 1524–1528

Davies CA, Mann DMA, Sumpter PQ, Yates PO (1987) A quantitative morphometric analysis of the neuronal and synaptic content of the frontal and temporal cortex in patients with Alzheimer's disease. J Neurol Sci 78: 151–164

DeKosky ST, Scheff SW (1990) Synapse loss in frontal cortex biopsies in Alzheimer's disease: Correlation with cognitive severity. Ann Neurol 27: 457–464

Delaere P, Duyckaerts C, Brion JP, Poulain V, Hauw JJ (1989) Tau, paired helical filaments and amyloid in the neocortex: a morphometric study of 15 cases with graded intellectual status in aging and senile dementia of Alzheimer type. Acta Neuropathol 77: 645–653

Dickson DW, Farlo J, Davies P, Crystal H, Fuld P, Yen SC (1988) Alzheimer disease. A double immunohistochemical study of senile plaques. Am J Pathol 132: 86–101

Dickson DW, Crystal HA, Bevona C, Honer W, Vincent I, Davies P (1995) Correlations of synaptic and pathological markers with cognition of the elderly. Neurobiol Aging 16: 285–304

Eccles JC (1981) The modular operation of the cerebral neocortex considered as the material basis of mental events. Neuroscience 6: 1839–1856

Eccles JC (1984) The cerebral neocortex: a theory of its operation. In: Jones EG, Peters A (eds) Cerebral cortex. Volume 2. Functional properties of cortical cells. Plenum Press, New York, pp 1–38

Games D, Adams D, Alessandrini R, Barbour R, Berthelette P, Blackwell C, Carr T, Clemes J, Donaldson T, Gillespie F, Guido T, Hagopian S, Johnson-Wood K, Khan K, Lee M, Leibowitz P, Lieberburg I, Little S, Masliah E, McConlogue L, Montoya-Zavala M, Mucke L, Paganini L, Penniman E, Power M, Schenk D, Seubert P, Snyder B, Soriano F, Tan H, Vitale J, Wadsworth S, Wolozin B, Zhao J (1995) Alzheimer-type neuropathology in transgenic mice overexpressing V717F β-amyloid precursor protein. Nature 373: 523–527

Goate A, Chartier-Harlin M-C, Mullan M, Brown J, Crawford F, Fidani L, Giuffra L, Haynes A, Irving N, James L, Mant R, Newton P, Rooke K, Roques P, Talbot C, Williamson R, Rossor M, Owen M, Hardy J (1991) Segregation of a missense mutation in the amyloid precursor protein gene with familial Alzheimer's disease. Nature 349: 704–706

Golde TE, Estus S, Younkin LH, Selkoe DJ, Younkin SG (1992) Processing of the amyloid protein precursor to potentially amyloidogenic derivatives. Science 255: 728–730

Gordon I, Grauer E, Genis I, Sehayek E, Michaelson DM (1995) Memory deficits and cholinergic impairments in apolipoprotein E-deficient mice. Neurosci Lett 199: 1–4

Greenamyre JT, Porter RHP (1994) Anatomy and physiology of glutamate in the CNS. Neurology 44 (suppl): S7–S13

Gregg RE, Zech LA, Schaefer EJ, Stark D, Wilson D, Brewer HB Jr (1986) Abnormal in vivo metabolism of apolipoprotein E$_4$ in humans. J Clin Invest 78: 815–821

Hamos JE, DeGennaro LJ, Drachman DA (1989) Synaptic loss in Alzheimer's disease and other dementias. Neurology 39: 355–361

Hartmann H, Eckert A, Muller WE (1994) Apolipoprotein E and cholesterol affect neuronal calcium signalling: the possible relationship to beta-amyloid neurotoxicity. Biochem Biophys Res Commun 200: 1185–1192

Heinonen O, Soininen H, Sorvari H, Kosunene O, Paljarvi L, Koivisto E, Riekkinen PJ (1995) Loss of synaptophysin-like immunoreactivity in the hippocampal formation is an early phenomenon in Alzheimer's disease. Neuroscience 64: 375–384

Hof PR, Morrison JH (1994) The cellular basis of cortical disconnection in Alzheimer disease and related dementing conditions. In: Terry RD, Katzman R, Bick KL (eds) Alzheimer disease. Raven Press, New York, pp 197–230

Hof PR, Cox K, Morrison JH (1990) Quantitative analysis of a vulnerable subset of pyramidal neurons in Alzheimer's disease: I. Superior frontal and inferior temporal cortex. J Comp Neurol 301: 44–54

Holtzman DM, Pitas RE, Kilbridge J, Nathan B, Mahley RW, Bu G, Schwartz AL (1995) Low density lipoprotein receptor-related protein mediates apolipoprotein E-dependent neurite outgrowth in a central nervous system-derived neuronal cell line. Proc Natl Acad Sci USA 92: 9480–9484

Honer WG, Dickson DW, Gleeson J, Davies P (1992) Regional synaptic pathology in Alzheimer's disease. Neurobiol Aging 13: 375–382

Hyman BT, Van Hoesen GW, Kromer LJ, Damasio AR (1986) Perforant pathway changes in the memory impairment of Alzheimer's disease. Ann Neurol 20: 472–481

Hyman BT, Kromer LJ, Van Hoesen GW (1987) Reinnervation of the hippocampal perforant pathway zone in Alzheimer's disease. Ann Neurol 21: 259–267

Ignatius MJ, Shooter EM, Pitas RE, Mahley RW (1987) Lipoprotein uptake by neuronal growth cones in vitro. Science 236: 959–962

Iwai A, Masliah E, Yoshimoto M, De Silva R, Ge N, Kittel A, Saitoh T (1994) The precursor protein of non-Aβ component of Alzheimer's disease amyloid (NACP) is a presynaptic protein of the central nervous system. Neuron 14: 467–475

Iwai A, Yoshimoto M, Masliah E, Saitoh T (1995) Non-Aβ component of Alzheimer's disease amyloid (NAC) is amyloidogenic. Biochemistry 34: 10139–10145

Iwai A, Masliah E, Sundsmo MP, DeTeresa R, Mallory M, Salmon DP, Saitoh T (1996) The synaptic protein NACP is abnormally expressed during the progression of Alzheimer's disease. Brain Res 720: 230–234

Jensen PH, Sorensen ES, Petersen TE, Gliemann J, Rasmussen LK (1995) Residues in the synuclein consensus motif of the alpha synuclein fragment, NAC, participate in transglutaminase-catalysed cross-liking to Alzheimer-disease amyloid beta A4 peptide. Biochem J 310: 91–94

Kanai Y, Hediger MA (1992) Primary structure and functional characterization of a high-affinity glutamate transporter. Nature 360: 467–471

Kanai Y, Smith CP, Hediger MA (1993) The elusive transporters with a high affinity for glutamate. Trends Neurosci 16: 359–365

Kanai Y, Stelzner M, Nubberg S, Khawaja S, Hebert SC, Smith CP, Hediger MA (1994) The neuronal and epithelial human high affinity glutamate transporter. J Biol Chem 269: 20599–20606

Kanai Y, Nussberger S, Romero MF, Boron WF, Hebert SC, Hediger MA (1995) Electrogenic properties of the epithelial and neuronal high affinity glutamate transporter. J Biol Chem 270: 16561–16568

Kounnas MZ, Moir RD, Rebeck GW, Bush AI, Argaves WS, Tanzi RE, Hyman BT, Strickland DK (1995) LDL receptor-related protein, a multifunctional apoE receptor, binds secreted β-amyloid precursor protein and mediates its degradation. Cell 82: 331–340

Kovacs DM, Fausett HJ, Page KJ, Kim TW, Moir RD, Merriam DE, Hollister RD, Hallmark OG, Mancini R, Felsenstein KM (1996) Alzheimer-associated presenilins 1 and 2: neuronal expression in brain and localization to intracellular membranes in mammalian cells. Nature Med 2: 224–229

Lassmann H, Weiler R, Fischer P, Bancher C, Jellinger K, Floor E, Danielczyk W, Seitelberger F, Winkler H (1992) Synaptic pathology in Alzheimer's disease: immunological data for markers of synaptic and large dense-core vesicles. Neuroscience 46: 1–8

Lin-Lee YC, Kao FT, Cheung P, Chan L (1985) Apolipoprotein E gene mapping and expression: localization of the structural gene to human chromosome 19 and expression of ApoE mRNA in lipoprotein- and non lipoprotein-producing tissues. Biochemistry 24: 3751–3756

Mahley RW, Innerarity TL (1983) Lipoprotein receptors and cholesterol homeostasis. Biochim Biophys Acta 737: 197–222

Manfras BJ, Rudert WA, Trucco M, Boehm BO (1994) Cloning and characterization of a glutamate transporter cDNA from human brain and pancreas. Bioch Biophys Acta 1195: 185–188

Maroteaux L, Scheller RH (1991) The rat brain synucleins; family of proteins transiently associated with neuronal membrane. Mol Brain Res 11: 335–343

Masliah E (1995a) Mechanisms of synaptic dysfunction in Alzheimer's disease. Histol Histopathol 10: 509–519

Masliah E (1995b) The natural evolution of the neurodegenerative alterations in Alzheimer's disease. Neurobiol Aging 16: 280–282

Masliah E, Terry R (1993) The role of synaptic proteins in the pathogenesis of disorders of the central nervous system. Brain Pathol 3: 77–85

Masliah E, Terry R (1994) The role of synaptic pathology in the mechanisms of dementia in Alzheimer's disease. Clin Neurosci 1: 192–198

Masliah E, Terry RD, DeTeresa RM, Hansen LA (1989) Immunohistochemical quantification of the synapse-related protein synaptophysin in Alzheimer disease. Neurosci Lett 103: 234–239

Masliah E, Hansen L, Albright T, Mallory M, Terry RD (1991a) Immunoelectron microscopic study of synaptic pathology in Alzheimer disease. Acta Neuropathol 81: 428–433

Masliah E, Mallory M, Hansen L, Alford M, Albright T, DeTeresa R, Terry RD, Baudier J, Saitoh T (1991b) Patterns of aberrant sprouting in Alzheimer disease. Neuron 6: 729–739

Masliah E, Terry RD, Alford M, DeTeresa RM, Hansen LA (1991c) Cortical and subcortical patterns of synaptophysin-like immunoreactivity in Alzheimer disease. Am J Pathol 138: 235–246

Masliah E, Ellisman M, Carragher B, Mallory M, Young S, Hansen L, DeTeresa R, Terry RD (1992a) Three-dimensional analysis of the relationship between synaptic pathology and neuropil threads in Alzheimer disease. J Neuropathol Exp Neurol 51: 404–414

Masliah E, Mallory M, Ge N, Saitoh T (1992b) Protein kinases and growth associated proteins in plaque formation in Alzheimer's disease. Rev Neurosci 3: 99–107

Masliah E, Mallory M, Hansen L, Alford M, DeTeresa R, Terry R, Baudier J, Saitoh T (1992c) Localization of amyloid precursor protein in GAP43-immunoreactive aberrant sprouting neurites in Alzheimer's disease. Brain Res 574: 312–316

Masliah E, Mallory M, Hansen L, Alford M, DeTeresa R, Terry R (1993a) An antibody against phosphorylated neurofilaments identifies a subset of damaged association axons in Alzheimer's disease. Am J Pathol 142: 871–882

Masliah E, Miller A, Terry RD (1993b) The synaptic organization of the neocortex in Alzheimer's disease. Medical Hypotheses 41: 334–340

Masliah E, Mallory M, Hansen L, DeTeresa R, Alford M, Terry R (1994) Synaptic and neuritic alterations during the progression of Alzheimer's disease. Neurosci Lett 174: 67–72

Masliah E, Iwai A, Mallory M, Ueda K, Saitoh T (1995a) Altered presynaptic protein NACP is associated with plaque formation and neurodegeneration in Alzheimer's disease. Am J Pathol 148: 201–210

Masliah E, Mallory M, Alford M, Ge N, Mucke L (1995b) Abnormal synaptic regeneration in hAPP695 transgenic and APOE knockout mice. In: Iqbal K, Mortimer JA, Winblad B, Wisniewski HM (eds) Research advances in Alzheimer's disease and related disorders. John Wiley & Sons Ltd, pp 405–414

Masliah E, Mallory M, Alfort M, Veinbergs I, Roses AD (1995c) ApoE role in maintaining the integrity of the aging central nervous system. In: Roses AD, Weisgraber KH, Christen Y (eds) Apolipoprotein E and Alzheimer's disease. Springer-Verlag, Heidelberg, pp 59–73

Masliah E, Mallory M, Ge N, Alford M, Veinbergs I, Roses AD (1995d) Neurodegeneration in the CNS of apoE-deficient mice. Exp Neurol 136: 107–122

Masliah E, Mallory M, Veinbergs I, Miller A, Samuel W (1996a) Alterations in apolipoprotein expression during aging and neurodegeneration. Prog Neurobiol 50: 493–503

Masliah E, Samuel W, Veinbergs I, Mallory M, Mante M, Saitoh T (1996b) Neurodegeneration and cognitive impairment in apoE-deficient mice is ameliorated by infusion of recombinant apoE. Neuroscience, in press

Masliah E, Alford M, De Teresa R, Mallory M, Hansen L (1996b) Deficient glutamate transport is associated with neurodegeneration in Alzheimer's Disease. Ann Neurol 40: 759–766

Masliah E, Westland CE, Rockenstein EM, Abraham CR, Mallory M, Veinberg I, Sheldon E, Mucke L (1997a) Amyloid precursor proteins protect neurons of transgenic nice against acute and chronic excitotoxic injuries *in vivo*. Neuroscience In press

Masters CL, Multhaup G, Simms G, Pottglesser J, Martins RN, Beyreuther K (1985) Neuronal origin of a cerebral amyloid: neurofibrillary tangles of Alzheimer's disease contain the same protein as the amyloid of plaque cores and blood vessels. EMBO J 4: 2757–2763

Mattson MP, Cheng B, Culwell AR, Esch FS, Lieberburg I, Rydel RE (1993a) Evidence for excitoprotective and intraneuronal calcium-regulating roles for secreted forms of the β-amyloid precursor protein. Neuron 10: 243–254

Mattson MP, Cheng B, Smith-Swintosky VL (1993b) Mechanisms of neurotrophic factor protection against calcium- and free radical- mediated excitotoxic injury: Implications for treating neurodegenerative dissorders. Exp Neurol 124: 89–95

Mattson MP, Tomaselli KJ, Rydel RE (1993c) Calcium-destabilizing and neurodegenerative effects of aggregated β-amyloid peptide are attenuated by basic FGF. Brain Res 621: 35–49

McKee AC, Kosik KS, Kowall NW (1991) Neuritic pathology and dementia in Alzheimer's disease. Ann Neurol 30: 156–165

Morris JC, Storandt M, McKeel DW Jr, Rubin EH, Price JL, Grant EA, Berg L (1996) Cerebral amyloid deposition and diffuse plaques in "normal" aging: Evidence for presymptomatic and very mild Alzheimer's disease. Neurology 46: 707–719

Morrison JH, Lewis DA, Campbell MJ (1987) Distribution of neurofibrillary tangles and nonphosphorylated neurofilament protein-immunoreactive neurons in cerebral cortex: implications for loss of corticocortical circuits in Alzheimer's disease. In: Davies P, Finch CE (eds) Molecular neuropathology of aging. Branbury Report, Vol. 27. Cold Springs Harbor Laboratory, New York, pp 109–124

Mucke L, Masliah E, Johnson WB, Ruppe MD, Rockenstein EM, Forss-Petter S, Pietropaolo M, Mallory M, Abraham CR (1994) Synaptotrophic effects of human amyloid β protein precursors in the cortex of transgenic mice. Brain Res 666: 151–167

Mucke L, Abraham CR, Ruppe MD, Rockenstein EM, Toggas SM, Alford M, Masliah E (1995) Protection against HIV-1 gp120-induced brain damage by neuronal overexpression of human amyloid precursor protein (hAPP). J Exp Med 181: 1551–1556

Nakajo S, Tsukada K, Omata K, Nakamura Y, Nakaya K (1993) A new brain-specific 14-kDA protein is a phosphoprotein. Its complete amino acid sequence and evidence for phosporylation. Eur J Biochem 217: 1057–1063

Nathan BP, Bellosta S, Sanan DA, Weisgraber KH, Mahley RW, Pitas RE (1994) Differential effects of apolipoproteins E3 and E4 on neuronal growth in vitro. Science 264: 850–852

Neary D, Snowdon JS, Mann DMA, Bowen DM, Sims NR, Northen B, Yates PO, Davison AN (1986) Alzheimer's disease: a correlative study. J Neurol Neurosurg Psych 49: 229–237

Paik Y-K, Chang DJ, Reardon CA, Davies GE, Mahley RW, Taylor JM (1985) Nucleotide sequence and structure of the human apolipoprotein E gene. Proc Natl Acad Sci USA 82: 3445–3449

Perry EK, Perry RH, Blessed G, Tomlinson BE (1977) Neurotransmitter enzyme abnormalities in senile dementia: CAT and GAD activities in necropsy tissue. J Neurol Sci 34: 247–265

Pines G, Danbolt NC, Bjoras M, Zhang Y, Bendahan A, Eide L, Koepsell H, Storm-Mathisen J, Seeberg E, Kanner BI (1992) Cloning and expression of a rat brain L-glutamate transporter. Nature 360: 464–467

Poirier J (1994) Apolipoprotein E in animal models of CNS injury and in Alzheimer's disease. Trends Neurosci 17: 525–530

Poirier J, Baccichet A, Dea D, Gauthier S (1993) Cholesterol synthesis and lipoprotein reuptake during synaptic remodeling in hippocampus in adult rats. Neurosci 55: 81–90

Poirier J, Delisle M-C, Quirion R, Aubert I, Farlow M, Lahiri D, Hui S, Bertrand P, Nalbantoglu J, Gilfix BM, Gauthier S (1995) Apolipoprotein E4 allele as a predictor of cholinergic deficits and treatment outcome in Alzheimer disease. Proc Natl Acad Sci USA 92: 12260–12264

Rebeck GW, Reiter JS, Strickland DK, Hyman BT (1993) Apolipoprotein E in sporadic Alzheimer's disease: allelic variation and receptor interactions. Neuron 11: 575–580

Roch J-M, Masliah E, Roch-Levecq A-C, Sundsmo MP, Otero DAC, Veinbergs I, Saitoh T (1994) Increase of synaptic density and memory retention by a peptide representing the trophic domain of the amyloid β/A4 protein precursor. Proc Natl Acad Sci USA 91: 7650–7654

Roses AD, Einstein G, Gilbert J, Goedert M, Han S-H, Huang D, Hulette C, Masliah E, Pericak-Vance MA, Saunders AM, Schmechel DE, Strittmatter WJ, Weisgraber KH, Xi P-T (1996) Morphological, biochemical, and genetic support for an apolipoprotein E effect on microtubular metabolism. Ann NY Acad Sci 777: 147–157

Rothstein JD, Jin L, Dykes Hoberg M, Kuncl RW (1993) Chronic inhibition of glutamate uptake produces a model of slow neurotoxicity. Proc Natl Acad Sci USA 90: 6591–6595

Rothstein JD, Martin L, Levey AI, Dykes-Hoberg M, Jin L, Wu D, Nash N, Kuncl RW (1994) Localization of neuronal and glial glutamate transporters. Neuron 13: 713–725

Rothstein JD, Van Kammen M, Levey AI, Martin LJ, Kuncl RW (1995) Selective loss of glial glutamate transporter GLT-1 in amyotrophic lateral sclerosis. Ann Neurol 38: 73–84

Saitoh T, Roch J-M, Lin LW, Ninomiya H, Otero DAC, Yamamoto K, Masliah E (1994) The biological function of amyloid β/A4 protein in precursor. In: Masters CL, Beyreuther K, Trillet M, Christen Y (eds) Amyloid protein precursor in development, aging and Alzheimer's disease. Springer-Verlag, Heidelberg, pp 90–99

Samuel W, Masliah E, Terry R (1994a) Hippocampal connectivity and Alzheimer's dementia: effects of pathology in a two-component model. Neurology 44: 2081–2088

Samuel W, Terry RD, DeTeresa R, Butters N, Masliah E (1994b) Clinical correlates of cortical and nucleus pathology in Alzheimer dementia. Arch Neurol 51: 772–778

Saunders AM, Strittmatter WJ, Schmechel D, St.George-Hyslop PH, Pericak-Vance MA, Joo SH, Rosi BL, Gusella JF, Crapper-MacLachlan DR, Alberts MJ, Hulette C, Crain B, Goldgaber D, Roses AD (1993) Association of apolipoprotein E allele E4 with late-onset familial and sporadic Alzheimer's disease. Neurology 43: 1467–1472

Schmechel DE, Saunders AM, Strittmatter WJ, Crain BJ, Hulette CM, Joo SH, Pericak-Vance MA, Goldgaber D, Roses AD (1993) Increased amyloid beta-peptide deposition in cerebral cortex as a consequence of apolipoprotein E genotype in late-onset Alzheimer disease. Proc Natl Acad Sci USA 90: 9649–9653

Schubert W, Prior R, Weidemann A, Dircksen H, Multhaup G, Masters CL, Beyreuther K (1991) Localization of Alzheimer beta A4 amyloid precursor protein at central and peripheral synaptic sites. Brain Res 563: 184–194

Scott HL, Tannenberg AEG, Dodd PR (1995) Variant forms of neuronal glutamate transporter sites in Alzheimer's disease cerebral cortex. J Neurochem 64: 2193–2202

Selkoe DJ (1989) Amyloid β protein precursor and the pathogenesis of Alzheimer's disease. Cell 58: 611–612

Selkoe DJ 1993) Physiological production of the β-amyloid protein and the mechanisms of Alzheimer's disease. Trends Neurosci 16: 403–409

Shashidharan P, Huntley GW, Meyer T, Morrison JH, Platakis A (1994) Neuron-specific human glutamate transporter: molecular cloning, characterization and expression in human brain. Brain Res 662: 245–250

Shimokawa M, Yanagisawa K, Nishiye H, Miyatake T (1993) Identification of amyloid precursor protein in synaptic plasma membrane. Biochem Biophys Res Commun 196: 240–244

Sisodia SS, Koo EH, Beyreuther K, Unterbeck A, Price DL (1990) Evidence that β-amyloid protein in Alzheimer's disease is not derived by normal processing. Science 248: 492–494

Small DH, Nurcombe V, Reed G, Clarris H, Moir R, Beyreuther K, Masters CL (1994) A heparin-binding domain in the amyloid protein precursor of Alzheimer's disease is involved in the regulation of neurite outgrowth. J Neurosci 14: 2117–2127

Smith-Swintosky VL, Pettigrew LC, Craddock SD, Culwell AR, Rydel RE, Mattson MP (1994) Secreted forms of β-amyloid precursor protein protect against ischemic brain injury. J Neurochem 63: 781–784

Storck T, Schulte S, Hofman K, Stoffel W (1992) Structure, expression, and functional analysis of a Na+-dependent glutamate/aspartate transporter from rat brain. Proc Natl Acad Sci USA 89: 10955–10959

Strickland DK, Kounas MZ, Argaves WS (1995) LDL receptor-related protein: a multiligand receptor for lipoprotein and proteinase catabolism. FASEB J 9: 890–898

Strittmatter WJ, Saunders AM, Schmechel D, Pericak-Vance M, Enghild J, Salvesen GS, Roses AD (1993) Apolipoprotein E: high-aviditiy binding to β-amyloid and increased frequency of type 4 allele in late-onset familial Alzheimer disease. Proc Natl Acad Sci USA 90: 1977–1981

Strittmatter WJ, Weisgraber KH, Goedert M, Saunders AM, Huang D, Corder EH, Dong LM, Jakes R, Alberts MJ, Gilbert JR (1994) Hypothesis: microtubule instability and paired helical filament formation in the Alzheimer disease brain are related to apolipoprotein E genotype. Exp Neurol 125: 163–171

Suzuki K, Terry RD (1967) Fine structural localization of acid phosphatase in senile plaques in Alzheimer's presenile dementia. Acta Neuropathol 8: 276–284

Tanzi RE, McClatchey AI, Lamperti ED, Villa-Komaroff L, Gusella JF, Neve RL (1988) Protease inhibitor domain encoded by an amyloid protein precursor mRNA associated with Alzheimer's disease. Nature 331: 528–530

Terry RD, Gonatas NK, Weiss M (1964) Ultrastructural studies in Alzheimer's presenile dementia. Am J Pathol 44: 269–297

Terry RD, Wisniewski HM (1970) The ultrastructure of the neurofibrillary tangle and the senile plaque. In: Wolstenholme GEW, O'Connor M (eds) Ciba Foundation Symposium on Alzheimer's Disease and Related Conditions. J & A Churchill, London, pp 145–168

Terry RD, Peck A, DeTeresa R, Schechter R, Horoupian DS (1981) Some morphometric aspects of the brain in senile dementia of the Alzheimer type. Ann Neurol 10: 184–192

Terry RD, Masliah E, Salmon DP, Butters N, DeTeresa R, Hill R, Hansen LA, Katzman R (1991) Physical basis of cognitive alterations in Alzheimer disease: synapse loss is the major correlate of cognitive impairment. Ann Neurol 30: 572–580

Ueda K, Fukushima H, Masliah E, Xia Y, Iwai A, Otero D, Kondo J, Ihara Y, Saitoh T (1993) Molecular cloning of a novel amyloid component in Alzheimer's disease. Proc Natl Acad Sci USA 90: 11282–11286

Ueda K, Saitoh T, Mori H (1994) Tissue-dependent alternative splicing of mRNA for NACP, the precursor of non-A beta component of Alzheimer's disease amyloid. Biochem Biophys Res Commun 205: 1366–1372

Van Nostrand WE, Wagner SL, Suzuki M, Choi BH, Farrow JS, Geddes JW, Cotman CW, Cunningham DD (1989) Protease nexin-II, a potent antichymotripsin, shows identity to amyloid β protein. Nature 341: 546–548

Xia Y, de Silva HAR, Rosi BL, Yamaoka LH, Rimmler JB, Pericak-Vance MA, Roses AD, Chen X, Masliah E, DeTeresa R, Iwai A, Sundsmo M, Thomas RG, Hofstetter CR, Gregory E, Hansen LA, Katzman R, Thal LJ, Saitoh T (1996) Genetic studies in Alzheimer's disease with an NACP/alpha-synuclein polymorphism. Ann Neurol 40: 207–215

Yamada T, Kondo A, Takamatsu J, Tateishi J, Goto I (1995) Apolipoprotein E mRNA in the brains of patients with Alzheimer's disease. J Neurol Sci 129: 56–61

Yamaguchi H, Hirai S, Morimatso M, Shoji M, Ihara Y (1988) A variety of cerebral amyloid deposits in the brains of Alzheimer-type dementia demonstrated by β-protein immunostaining. Acta Neuropathol 76: 541–549

Yoshimoto M, Iwai A, Kang D, Otero DAC, Xia Y, Saitoh T (1995) NACP, the precursor protein of non-amyloid β/A4 protein (Aβ) component of Alzheimer disease amyloid, binds Aβ and stimulates Aβ aggregation. Proc Natl Acad Sci USA 92: 9141–9145

Zannis VL, Breslow JL (1981) Human very low density lipoprotein apolipoprotein E isoprotein polymorphism is explained by genetic variations and post-translational modification. Biochemistry 20: 1033–1041

Zhan SS, Beyreuther K, Schmitt HP (1993) Quantitative assessment of the synaptophysin immunoreactivity of the cortical neuropil in various neurodegenerative disorders with dementia. Dementia 4: 66–74

Zhong Z, Quon D, Higgins LS, Higaki J, Cordell B (1994) Increased amyloid production from aberrant β-amyloid precursor proteins. J Biol Chem 269: 12179–12184

Is Alzheimer's Disease Accelerated Aging? Different Patterns of Age and Alzheimer's Disease-Related Neuronal Losses in the Hippocampus

M. J. West[1], P. D. Coleman, D. G. Flood, and J. C. Troncoso

Summary

The hypothesis that Alzheimer's disease is accelerated aging has been tested by comparing the patterns of neuron loss in the hippocampal region associated with normal aging and with Alzheimer's disease. Qualitative differences were found in the patterns of loss in CA1, indicating that a unique neurodegenerative process not associated with normal aging is involved in Alzheimer's disease.

Introduction

The question raised in the title of this paper underscores our continued lack of understanding of the relationship between aging and Alzheimer's disease (AD); Berg 1985). The increased frequency of AD with age (Bachman et al. 1993) and the presence of senile plaques (SP) and neurofibrillary tangles (NFT) in cognitively intact aged individuals (Katzman et al. 1988) suggest that the pathological changes associated with AD and with normal aging represent a continuum. These observations further suggest that the neurodegenerative processes and mechanisms associated with normal aging and AD are similar and the distinction between the two is quantitative in nature. In brief, if we live long enough, we will all get AD. The international criteria used to make the neuropathological diagnosis of AD implicitly support this position. Invariably, these criteria (e.g., Khachaturian 1985) involve age-adjusted threshold levels of SPs and NFTs in specific parts of the brain and are based on quantitative distinctions.

On the other hand, the identification of environmental risk factors such as trauma, metals, and viruses, and genetic risk factors such as amyloid precursor proteins (Hardy et al. 1991) and apolipoproteins (Corder et al. 1993) suggests that AD involves neurodegenerative mechanisms that do not characterize normal aging and that it is more appropriate to view AD as a true disease. Evidence of rapid, catastrophic atrophic events in the medial temporal lobe of longitudinally studied cases of probable AD further supports this view (Jobst et al. 1994). the resolution of the issue as to whether AD is an accelerated form of aging or a sepa-

---

[1] Department of neurobiology, Institute of Anatomy, University of Aarhus, 8000 Aarhus C, Denmark

B. T. Hyman / C. Duyckaerts / Y. Christen (Eds.)
Connections, Cognition, and Alzheimer's Disease
© Springer-Verlag Berlin Heidelberg 1997

rate pathological process that warrants the distinction of a true disease has important implications for the design of investigative, therapeutic, and preventative strategies. For if AD involves the same neurodegenerative mechanisms as those associated with normal aging, it would be highly appropriate to develop strategies that focus on a better understanding of normal aging and how it can be modulated. If evidence can be obtained to support the position that AD is a true disease, then the focus should be placed on the identification and modulation of a unique process not associated with aging.

What sort of anatomical evidence would allow one to distinguish between these two alternatives? First, it must involve a parameter that will potentially permit a distinction to be made between quantitative and qualitative differences in the brains of AD and normal aging subjects. This eliminates SPs and NFTs. Both are structures that most likely develop and disappear during the course of AD and normal aging and their numbers at any one point in time may not necessarily reflect the extent of the neurodegenerative process. More importantly, they can never be expected to provide evidence of qualitative differences because they are present to certain degrees in the brains of both AD and normal aging subjects. Neuron number or, better said, neuron loss would be a more suitable parameter in view of the fact that neurons are not generated in the mature nervous system and neuron loss can be expected to be reflective of the cumulative damage that has taken place in specific parts of the brain. Evaluations of neuron loss associated wit AD and normal aging thus have the potential for distinguishing between quantitative and qualitative changes in the brains of AD and normal aged subjects. In the former, this would appear as different degrees of loss, in the latter, it would appear as a loss in the AD subjects in a subdivision that did not show evidence of age-related loss. Second, though not essential for the formulation of a hypothesis, this parameter should be investigated in a part of the brain believed to be involved in the primary pathological events associated with AD. Ideally this analysis would also include a number of readily definable neuron types, some but not all of which demonstrate normal age-related loss.

To test the hypothesis that AD is accelerated aging, we have evaluated normal age-related and AD-related neuron loss in five of the major subdivisions of the hippocampal region of the brain. The hippocampal region was chosen because it is one of the first regions to develop SPs and NFTs, remains one of the most profoundly affected regions during the course of AD, and is known to be involved in memory processes that become defective during the early clinical phases of the disease (Hyman et al. 1990). In addition, it comprises a number of well-defined subdivisions with distinctive neuronal types. Support of the hypothesis would require evidence of different degrees of loss of the same types of neurons. Evidence of either AD- or age-related loss, but not both, in one or more subdivisions would result in rejection of the hypothesis and constitute evidence of qualitative differences in the pathological processes involved.

Materials and Methods

Subjects and Comparative Methods

Estimates of the total number of neurons in the hippocampal subdivisions were made on the left side of the brain of two groups of male subjects (West et al. 1994). One group, the normal aging group (NAG), comprised 38 subjects who had no history of mental illness, neurological disease, or long-term illness (mean age 56 years, range 13–101). Normal age-related neuronal loss was evaluated for each subdivision by testing the regression of total neuron number with age for all the individuals in the NAG. Regressions with 2p values of less than 5 % were considered significant. For regions in which there was a significant negative relationship, normal age-related neuron loss was expressed in terms of the ratios of the regression values at the youngest and oldest ages represented in the NAG.

The second group consisted of seven subjects (mean age 79 years, range 68–88), who, on the basis of clinical and pathological criteria, were diagnosed as moderate to severe cases of AD (ADG). For comparative purposes, a third, age-matched control group (AMCG) was formed from 14 individuals in the NAG (mean age 78 years, range 64–101). AD-related neuronal loss was evaluated for each subdivision with unpaired t-tests performed on the data from the ADG and the AMCG. In these comparisons, 2p values of 5 % or less were considered significant.

Estimates of Total Neuron Number

Unbiased estimates were made of the total number of neurons in each of the following subdivisions of the hippocampal region of each individual: dentate granule cells (GRAN), dentate hilar cells (HIL), the combined pyramidal cell layers of CA3 and CA2 (CA3–2) and the pyramidal cell layer of CA1 (CA1), and cellular layers of subiculum (SUB; Fig. 1). The individual estimates were obtained by multiplying the volume of the layer containing the cell bodies (obtained by point counting techniques) by the numerical density of the neurons in those layers (obtained with optical disectors) (West and Gundersen 1990). For the analysis, 12 to 15 sections, chosen in a systematic random manner at equal intervals along the entire extent of the hippocampal region, were used to ensure that all parts of the subdivision under consideration had equal probabilities of being sampled. The sampling schemes were designed so that the true inter-individual variance, not the precision of the individual estimates, made the major contribution to the group variance.

Fig. 1. Histological section through the medial temporal lobe of the brain (top) and a corresponding diagrammatic representation (bottom) of the subdivisions of the hippocampal region in which age-related and AD-related neuronal losses were evaluated. SUB, subiculum; HIL, dentate hilar cells; GRAN, dentate granular cells.

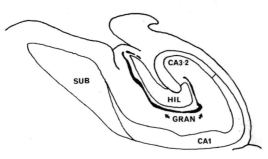

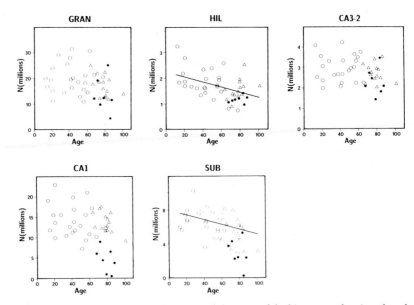

Fig. 2. The total number of neurons (N) in the major subdivisions of the hippocampal region plotted as a function of age. Open circles and open triangles, the normal aging group, (NAG; n = 38); filled circles, the AD group (ADG; n = 7); open triangles, the age-matched control group (AMCG; n = 14). Abbreviations same as for Figure 1.

Results

Estimates of the total number of neurons in each subdivision of each individual are shown graphically in Figure 2. In the NAG, there were significant negative correlations between age and neuron number in HIL and SUB. The regression lines for these two subdivisions are shown in Figure 2 and indicate losses of 37 % in HIL and 43 % in SUB over the ages studied.

When the group means of total number of neurons in the various subdivisions of the ADG and the AMCG were compared, significantly lower values were found in HIL, CA1, and SUB, representing AD-related neuronal losses of 25 %, 68 % and 47 %, respectively.

The percentage of neurons lost in the various subdivisions that were age-related and AD-related are shown in Figure 3. In summary, there were neither age-related nor AD-related neuron loss in GRAN or CA3. In HIL there was an age-related loss (37 %) and an additional AD-related loss (25 %). In SUB there was also an age-related loss (43 %) and an additional AD-related loss (47 %). Notably, there was no age-related loss in CA1, but there was evidence a massive loss related to AD (68 %).

Discussion

The regional patterns of neuronal loss in the hippocampus related to normal aging and AD are qualitatively different. This conclusion is based on the absence of significant age-related changes in the number of neurons in CA1 and the presence of a significant AD-related losses in this subdivision. The presence of both age-related and AD-related neuronal loss in HIL and SUB indicates that the dis-

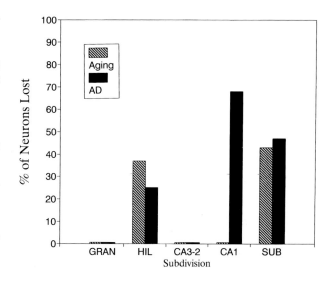

Fig. 3. The percentage of neurons lost due to aging (striped bars) and AD (solid bars). The aging losses were calculated from the end points of the regression lines of age versus neuron number for subdivisions in which there were significant correlations (HIL, SUB). The AD losses were calculated from the means of the numbers of neurons in the ADG and AMCG for subdivisions in which there were significant differences in the group means. Abbreviations same as for Figure 1.

tinction between age-related and AD-related neuronal losses is quantitative in nature and would be compatible with the hypothesis that AD is accelerated aging. However, in order to explain the absence of age-related losses and the presence of AD losses in CA1, it is necessary to involve a neurodegenerative process in AD that is not present during normal aging. This constitutes evidence of a qualitative difference in the neurodegenerative processes associated with normal aging and with AD and supports the alternative hypotheses that AD is a true disease process and involves a unique neurodegenerative process not associated with normal aging.

Two alternative interpretations of these findings must be considered. The first is that secular differences are involved. Because the material used in this study was collected over the span of a few years, i.e., it is a cross-sectional study, it is possible that the age-related changes and the differences between the ADG and the AMCG were present at an early age and do not represent neuronal loss. The other possibility is that differences in the agonal states of the individual in the ADG and the AMCG may account for the differences in the numbers of CA1 neurons in these two groups. It is well established that global ischemia can result in cell-specific neurodegenerative events in CA1 while adjacent regions appear unaffected (Zola-Morgan et al. 1986). Accordingly a chronic ischemic premortem state, disproportionately represented in the ADG and not directly related to the disease process, may be involved in the selective loss of neurons observed in CA1 in this group. The longitudinal study of Jobst et al. (1994), describing rapid progressive atrophy in a part of the medial temporal lobe of ambulatory AD patients that can be demonstrated to correspondent primarily to CA1, argues against both of these interpretations.

The conclusion that there are qualitative differences in the neurodegenerative processes associated with normal aging and AD has several implications. Perhaps the most important of these is that AD is not an inevitable consequence of aging. It should be considered to be a true disease characterized by a unique neurodegenerative process that cannot be thoroughly understood through studies of normal aging. Because this process is not confounded by age-related neurodegenerative processes in CA1, its identification can be expected more readily carried out in this subdivision of the hippocampal region.

In view of the incontrovertible evidence of numerous studies that there is an increased frequency of AD with age and the evidence presented here that separate neurodegenerative processes are involved in AD and normal aging, the question arises as to the relationship between age and AD. The most obvious answer to this is time. If AD is not directly related to the aging process and its incidence increases with age or time, then AD must involve a process that takes years if not decades to manifest itself. The relatively short longevity of most other species may be the prime reason why AD has not been identified in other species. According to this scenario the causative insult may occur long before the symptoms of AD appear. In the interim period slow changes, masked by compensatory responses from remaining components of the region affected or other parts of the nervous system, would take place until the point at which the compensatory reserves are depleted to the extent that the functional integrity of specific systems is compromised. The end point of this process would be the collapse of specific regions of the brain and the relatively acute onset of symptoms. Such a scenario suggests that it would be appropriate to search for diagnostic indices long before the clinical signs of AD appear.

Acknowledgments

Supported by the NIA-funded Alzheimer's Disease Research Centers at The University of Rochester and Johns Hopkins University. A preliminary report of these findings was published in The Lancet 344: 769–772, 1994.

References

Bachman DL, Wolf PA, Linn RT, Knoefel JE, Cobb JL, Belanger AJ, White LR, D'Agostino RB (1993) Incidence of dementia and probable Alzheimer's disease in a general population: The Framingham Study. Neurology 43: 515–519

Berg L (1985) Does Alzheimer's disease represent an exaggeration of normal aging? Arch Neurol 42: 737–739

Corder EH, Saunders AM, Strittmatter WJ, Schmechel DE, Gaskell PC, Small GW, Roses AD, Haines JL, Pericak-Vance MA (1993) Gene dose of apolipoprotein E type 4 allele and the risk of Alzheimer's disease in late onset families. Science 261: 921–923

Hardy J, Mullan M, Chartier-Harlin MC (1991) Molecular classification of Alzheimer's disease. Lancet 337: 1342

Hyman BT, Van Hoesen GW, Damasio AR (1990) Memory-related neural systems in Alzheimer's disease: an anatomic study. Neurology 40: 1721–1730

Jobst KA, Smith AD, Szatmari M, Esiri MM, Jaskowski A, Hindley N, McDonald B, Molyneux AJ (1994) Rapidly progressing atrophy of medial temporal lobe in Alzheimer's disease. Lancet 343: 829–830

Katzman R, Terry R, DeTeresa R, Brown T, Davies P, Fuld P, Reubing X, Peck A (1988) Clinical, pathological, and neurochemical changes in dementia: a subgroup with preserved mental status and numerous neocortical plaques. Ann Neurol 23: 138–144

Khachaturian ZS (1985) Diagnosis of Alzheimer's disease. Arch Neurol 42: 1097–1105

West MJ, Gundersen HJG (1990) Unbiased stereological estimation of the number of neurons in the human hippocampus. J Compar Neurol 296: 1–22

West MJ, Coleman PD, Flood DG, Troncoso JC (1994) Differences in the pattern of hippocampal neuronal loss in normal aging and Alzheimer's disease. Lancet 344: 769–772

Zola-Morgan S, Squire LR, Amaral DG (1986) Human amnesia and the medial temporal region: Enduring memory impairment following a bilateral lesion limited to field CA1 of the hippocampus. J Neurosci 6: 2950–2967

Connections and Cognitive Impairment in Alzheimer's Disease

T. Gomez-Isla * and B. T. Hyman*

The clinical hallmark of Alzheimer's disease (AD) is dementia, that is, a stereotyped syndrome of cognitive dysfunction dominated by impairment of memory functions, including delayed recall, and ultimately affecting a wide range of higher cognitive functions. The neuropathological hallmarks of AD were described in Alzheimer's initial report in 1906 and include neurofibrillary tangles (NFT) and senile plaques (SP). The fundamental relationship between these lesions and the problem of what underlies the clinical syndrome has remained unsolved since the beginning of the century. Are NFT and SP directly related to clinical symptoms, or are they epiphenomena that signal synaptic and neuronal loss? In the last decade, these questions have been revisited through systematic studies of individuals with various degrees of dementia using newer immunohistochemical techniques and quantitative approaches to neuropathological study. Combined with a greater appreciation for the connectivity of the primate brain derived from neuroanatomical observations from the nonhuman primate, these new studies have come together to provide a relatively uniform picture of AD dementia as resulting from collapse of specific neural systems, as pathological changes selectively and specifically involve structures related to memory and higher cognitive function. These observations provide satisfying clinical-pathological answers and give birth to the next generation of questions: why are these neural populations vulnerable? and how do these patterns of vulnerability fit into the pictures generated by new information on genetic predisposition and causes of the disease? Answers to these new questions will hopefully lead to insight into basic pathophysiological mechanisms in the disease.

Patterns of hierarchical vulnerability suggest that disruption of memory-related neural systems underlies memory impairment in Alzheimer's disease

Neuropathological study of the AD brain demonstrates that NFT and SP do not appear in a random or widespread uniform fashion but instead follow a remarkably selective distribution (Fig. 1a and 1b from Arnold et al. 1991; Braak and

* Neurology Service, Massachusetts General Hospital, Fruit Street, Boston, MA 02114, USA

B. T. Hyman / C. Duyckaerts / Y. Christen (Eds.)
Connections, Cognition, and Alzheimer's Disease
© Springer-Verlag Berlin Heidelberg 1997

Braak 1991). This distribution is consistent across the majority of patients with AD and has been used as the basis for Braak's pathological staging of the disease (Braak and Braak 1991). NFT and SP are differentially distributed: the topography of NFT is far more consistent than that of SP and parallels the clinical symptoms of the disease more closely (Ball 1978; Hyman et al. 1984; Arnold et al. 1991; Braak and Braak 1991; Barnes and Pandya 1992; Hof et al. 1992; Arriagada et al. 1992a; Berg et al. 1993; Bouras et al. 1994). For example, NFT tend to occur in large projection neurons that are believed to be responsible for limbic-limbic and limbic-cortical connections. Perhaps the most consistent and severe alteration in the Alzheimer brain is the development of NFT in a subset of neurons in the medial temporal lobe (Hyman et al. 1984; Hof et al. 1992). The hierarchical vulnerability is reflected in the consistent order in which NFT affect cytoarchitectural fields. Among the first lamina and regions to be affected are neurons in layers II and IV of entorhinal cortex, and to a lesser extent also the perirhinal cortex, CA1/subiculum of the hippocampus, inferior temporal gyrus, amygdala and posterior parahippocampal gyri, cholinergic basal forebrain, and dorsal raphe (Arriagada et al. 1992a). Subsequently affected are, in order, high order association cortices, first order association cortices and ultimately other subcortical structures and primary sensory and motor cortices. We have postulated that these lesions disrupt projections between and among limbic and association cortices and especially affect the hippocampal formation. Destruction of these projections no doubt contributes to impaired cognition.

The first changes observed affect the primary afferent and efferent projections of the hippocampal formation, a structure central to normal memory function. NFT affect layers II and III of the entorhinal cortex and perirhinal cortex (Hyman et al. 1984; Mann and Esiri 1989; Braak and Braak 1991; Van Hoesen and Solodkin 1994), including the projection neurons within the superficial cell layers of entorhinal and perirhinal cortex which give rise to the perforant pathway. The perforant pathway is the final common projection for essentially all cortically derived information to the hippocampus; destruction of these entorhinal and perirhinal projections deafferents the hippocampus. SP frequently occur in the outer portion of the molecular layer of the dentate gyrus and in the stratum moleculare of the CA1/subicular field, the terminal zones of the perforant pathway (Hyman et al. 1986). The output of the hippocampal formation is similarly affected early in the disease process by marked accumulation of NFT in the CA1/subicular subfield (Ball 1978; Hyman et al. 1984). Two principal targets of the CA1/subicular projection are the accessory basal nucleus of the amygdala and layer IV of entorhinal cortex, which in turn broadcast hippocampal output to widespread cortical and subcortical targets. Both the accessory basal nucleus of the amygdala and layer IV of entorhinal cortex are also selectively and severely affected by NFT, and the accessory basal nucleus of the amygdala also frequently contains SP (Hyman et al. 1990). We have suggested that the integrity of neural systems related to memory is compromised by the loss of projections between and among the entorhinal cortex, hippocampal formation, and amygdala, taken together with NFT in the nucleus basalis, dorsal raphe, midline thalamic struc-

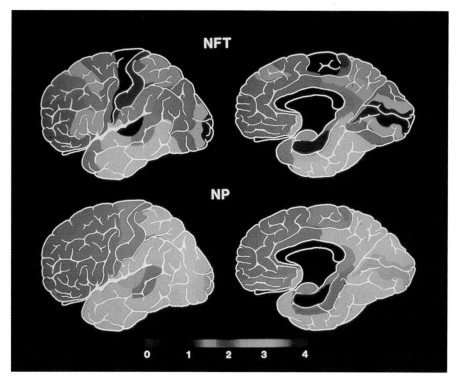

Fig. 1. A color imaging of the topography of NFT and neuritic plaques (NP) as assessed in 17 hemispheres using thioflavine S histochemistry. The color bar shows red as the greatest density and blue as the least (from Arnold et al. 1991).

tures, and paralimbic cortices such as the temporal pole (Hyman et al. 1990). Although there is no doubt a good deal of redundancy in these neural systems, we predicted that continued accumulation of NFT and of neuronal loss would ultimately lead to memory impairment. We have recently had the opportunity to test this postulate in a series of prospectively studied patients, as noted below.

As the disease progresses and the cognitive deficits become much broader, NFT can also be found in increasing numbers in feedforward and feedback projection neurons in layers II, III and V of high order association cortices (Lewis et al. 1987; Hof et al. 1990; Arnold et al. 1991). The temporal lobe, including inferior temporal gyrus and superior temporal sulcal area, appear to be affected earlier and more severely than parietal and frontal association cortices. Motor and sensory cortices remain unaffected by NFT even late in the disease process (Fig. 1a; Arnold et al. 1991; Price et al. 1991; Arriagada et al. 1992a). These anatomical observations emphasize the role of NFT in disrupting neural systems in AD brains.

Neurofibrillary tangles, but not senile plaques, parallel severity and duration of dementia

In contrast to the well-demarcated laminar and cytoarchitectural fields affected by NFT, SP have a broader and more varied distribution (Fig. 1 b). Although they also affect association cortices, SP may be present in primary motor and sensory cortices that remain clinically silent, even at the very late stages of the disease (Arnold et al. 1991; Arriagada et al. 1992a). SP do, however, tend to obey certain anatomical rules. For example, SP are often found in a discrete band in the outer portion of the molecular layer of the dentate gyrus, in an area which occupies the perforant pathway terminal zone (Hyman et al. 1986). In the amygdala, certain nuclei frequently contain Aβ, whereas others are relatively spared (Hyman et al. 1990). It is striking that many of these anatomic rules are also followed for Aβ deposition in the hAAPV717F transgenic mouse model of Aβ deposition, suggesting that unique anatomic considerations are critically important in defining the ultimate type and location of Aβ deposits (Games et al. 1995) We discovered some surprises when we studied whether the amount of SP pathology (defined as "amyloid burden, the percent of cortical surface area covered by Aβ") reflected clinical features. We used quantitative methods (computer image analysis techniques) to assess the amount of Aβ deposits in SP in individuals with various durations or severity of disease. In contrast to our data showing that NFT accumulate in parallel with severity of clinical illness, we were unable to find a correlation of SP with clinical symptoms (Arriagada et al. 1992a; Hyman et al. 1993; Gómez-Isla et al. 1996a). In fact, the number of SP in the cerebral cortex tends to remain stable independently of duration of illness, whereas the number of NFT correlates more closely with the severity of dementia (Arriagada et al. 1992a; Berg et al. 1993; Hyman et al. 1993; Hyman et al. 1995; Nagy et al. 1995; Gómez-Isla et al. 1996a). From this perspective, the presence of NFT qualitatively and quantitatively matches the clinical deficits of the disease better than the number or distribution of SP.

Neuronal Loss: Final Path to Alzheimer Dementia

Quantitative Assessment of Neuronal Loss, NFT, and Plaque Burden

By contrast to NFT and SP, neuronal and synaptic losses are negative phenomena that cannot easily be quantitated but only calculated from normative data on age-matched control populations. Nonetheless, it is obvious by inspection that the cortex of the Alzheimer brain is severely atrophied, and that there is substantial neural loss in many cytoarchitectural fields (Terry et al. 1981). Synaptic loss has also been clearly demonstrated (De Kosky and Scheff 1990; Terry et al. 1991). The cause of neuronal loss in neurodegenerative processes is unknown, and there are no biochemical markers to identify neurons at risk for death. As a consequence, no topographic map has been established to date to evaluate whether or not, and to what extent, there are consistent patterns of neuronal loss in AD brains. Does neuronal loss parallel tangles, plaques, both or neither?

The application of stereological counting techniques to the problem of counting neurons has provided a powerful tool to address these questions, overcoming the problems of correcting for cortical atrophy and adjusting appropriately for cell shrinkage and different cell sizes that lead to potential errors in traditional two-dimensional counting schemes (Coggeshall 1992; West 1993).

We have applied these new techniques to analyze the structural integrity of two regions of interest in AD that are consistently affected by both NFT and SP (Hyman et al. 1986, 1990; Arnold et al. 1991; Arriagada et al. 1992a; Hyman and Gómez-Isla 1994; Gómez-Isla et al. 1996): 1) the entorhinal cortex, a critical component of the medial temporal lobe memory-related neural system, and 2) the superior temporal sulcus (STS), a representation of high order association cortex. The STS is one of only three association areas in the non-human primate brain that receives input from each sensory modality (Pandya and Yeterian 1985; Barnes and Pandya 1992; Van Hoesen 1993).

Profound Neuronal Loss of Layer II Entorhinal Cortex Neurons Occurs in Very Mild Alzheimer's Disease

Our neuropathological studies of AD suggested that layer II of EC is selectively and severely affected by NFT. However, it is also quite common to find some NFT in layer II of EC even in (presumed) nondemented elderly individuals (Arriagada 1992b; Bouras et al. 1993). These observations led us to postulate that the AD-like neuropathological changes sometimes observed in nondemented elderly were in fact evidence of a presymptomatic state (Arriagada et al. 1992b). Similar conclusions were drawn by Braak, who assigned "stage I" to brains that contained NFT only in entorhinal/perirhinal cortices (Braak and Braak 1991). We recently had the opportunity to explore these issues in more detail by examining brains from elderly individuals who were known to be either cognitively normal or to have very mild symptoms of AD. We predicted that in truly normal aging there would be no or minimal neuronal loss, whereas even in early symptomatic AD neuronal loss in EC may be substantial.

We studied 20 individuals evaluated clinically in the Washington University Memory and Aging Project and the Washington University Alzheimer's Disease Research Center (ADRC) in collaboration with Drs. Joseph Price, Daniel McKeel, and John Morris at Washington University (Gómez-Isla et al. 1996a). From the clinical assessments, the 20 subjects were subdivided into two groups. The AD group included 10 clinically demented individuals (mean age $\pm$ SD = 84.2 $\pm$ 9.9 years; range, 67–95 years) rated according to the Clinical Dementia Rating (CDR) scale. Of the 10, 4 were rated CDR = 0.5 (very mild or questionable impairment), 1 CDR = 1 (mild) and 5 CDR = 3 (severe cognitive impairment). All

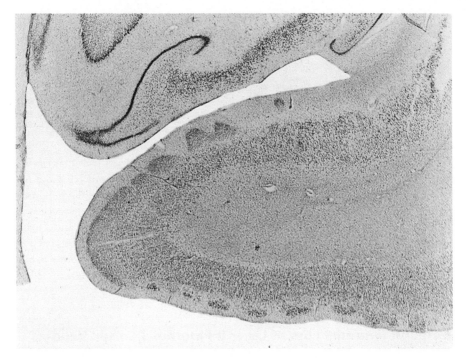

Fig. 2. Photomicrograph of Nissl-stained coronal section at the level of the uncus in a normal 71-year-old individual. Note the prominent cluster of neurons in the most superficial neuronal layer, layer II. These neurons are exquisitely vulnerable in AD.

subjects, including those with very mild clinical impairment, had a subsequent neuropathological diagnosis of definite AD (Khachaturian 1985; Mirra et al. 1991). The control group contained 10 individuals (mean age ± SD = 75 ± 9 years; range, 60–89 years) who clinically were rated CDR = 0 (cognitively normal). None of them met criteria for AD by neuropathological examination.

Lamina-specific neuronal counts were carried out following a systematically random sampling scheme on Nissl-stained sections representing the entire EC (Fig. 2) (West 1993). An average of 1,500 optical directors were assessed in each case.

We found that, in the cognitively normal (CDR = 0) individuals, there were about 650000 neurons in layer II, 1 million neurons in layer IV and 7 million neurons in the entire EC. The number of neurons remained constant between 60 and 90 years of age. The individuals with the mildest clinically detectable dementia (CDR = 0.5), all of whom had sufficient neurofibrillary tangles and senile plaques for the neuropathological diagnosis of AD, had 32 % fewer EC neurons than controls. The group with severe AD (CDR = 3) had 69 % fewer EC neurons than controls (Fig. 3).

Decreases in individual lamina in very mild AD cases (CDR = 0.5) were even more dramatic, with the number of neurons in layer II decreasing by 60 % and in

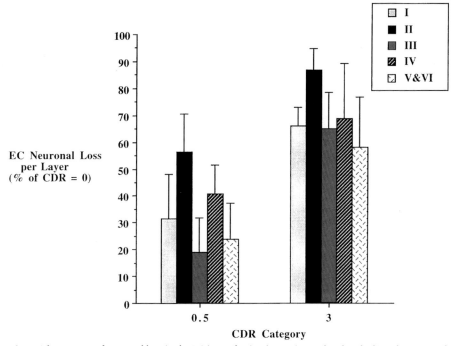

Fig. 3. The amount of neuronal loss in the EC in AD brains (n = 10) correlated with clinical severity of dementia in all layers. Layers II and IV, in particular, showed the most striking changes. In layer II, the degree of loss in the very mildly impaired CDR = 0.5 individuals was estimated to be 57 % compared to non-demented individuals. This increased to 87 % in the CDR = 3 individuals. (From Gómez-Isla et al. 1996a). These changes were each significant at the p < 0.05 level.

layer IV by 40 % compared to controls. In the severe dementia cases (CDR = 3), the number of neurons in layer II decreased by about 90 %, and the number of neurons in layer IV decreased by about 70 % compared to controls (Fig. 3).

These observations highlight the very severe neuronal loss in the EC even in very mild AD cases that are at the threshold for clinical detection of dementia. This neuronal loss is so marked that it must have started well before onset of clinical symptoms. In fact, if we assume a linear rate of loss, our data suggest that neuronal loss in layer II of EC may begin as long as seven to ten years prior to detectable clinical symptoms. This pattern may be similar to Parkinson's disease, wherein substantial loss of the substantia nigra has occurred by the time there are clinical symptoms. It is likely that a threshold of neuronal loss in the EC (along with accompanying alterations in other memory-related structures) must be reached before memory impairment becomes clinically apparent. The most dramatic neuronal loss selectively targets layers II and IV of EC, paralleling the known susceptibility of cells in these layers for NFT formation.

Relationship of Neuronal Loss to Synaptic and Neuropil Alterations

We have not directly assessed synaptic density or number in the EC or hippo-campal formation in these cases. However, the neuronal loss in the entire EC is accompanied by a parallel reduction of its volume, pointing to an associated loss of the neuropil, presumably including synapses. There is a close correlation between loss of synapses, loss of synaptophysin immunostaining, and cognitive deterioration (Terry et al. 1991), and perhaps the neuronal loss that we and West's studies (West et al. 1994) highlight contribute to the synaptic alterations that have been observed in the hippocampal formation and perforant pathway terminal zone (Masliah et al. 1991; Cabalka et al. 1992).

Neuronal Changes in EC Parallel NFT but not Amyloid SP

We also have compared EC neuron number to NFT and SP densities in the same 20 individuals from immunostained and Bielschowsky-stained adjacent sections. In general, NFT were predominant in layers II and IV with fewer numbers in III, V and VI, whereas SP were more scattered although amyloid plaques tended to occur in layer III. A significant negative correlation between the number of neu-rons and the degree of neurofibrillary tangles or neuritic plaques (which were highly correlated with each other) was observed, but no significant correlation was found between either of these two parameters and the number of total senile plaques or diffuse plaque subpopulations in the same area. These observations lead to the idea that neuronal loss and NFT accumulation in the EC, both signs of primary neuronal lesions, represent the most likely proximal correlates to the early memory impairment in AD, whereas a simple and direct correlation between Aβ deposition and neurotoxicity is unlikely.

Neuronal Loss in High Order Association Cortex Occurs at a Later Point in the Disease and Parallels the Chronological Evolution of Dementia

To assess the structural integrity of the association cortices at various point in the disease, we carried out a second study in which we counted the number of neurons in the STS in brains of 45 individuals with definite AD (average duration of illness, 8.2 ± 5.2 yrs, range, 0.3–21 yrs) and 28 non-demented control subjects (average age, 71.8 ± 14.8 yrs, range from the sixth to the tenth decade; Fig. 4). The 45 cases of AD selected for study met the following inclusion and exclusion criteria. They had been examined and followed in the clinical units of the Massa-chusetts General Hospital, Washington University or Mayo Clinic, had clear clin-ical histories of AD dementia with well-documented duration of disease, and had a neuropathological diagnosis of definite AD (Khachaturian 1985; Mirra et al. 1991). Using similar, statistically unbiased stereological counting techniques, approximately 1000 neurons were counted in each brain. Because of technical concerns regarding the identification of the anterior and posterior boundaries of the STS, we counted the number of STS neurons in a single 50 micron-thick slice rather than obtaining counts throughout the entire structure (as had been the case in the EC study). We tested whether neuropathological changes and neuronal loss were correlated with duration of illness as well as severity of dementia. Duration was assessed by determining, at first clinical examination, when the patient was "last normal." We charted measures of cognitive impairment derived from the information, memory and concentration (ICM) subtest of the Blessed Dementia Scale (BDS; Blessed et al. 1968). BDS scores within 18 months from death were available in nine AD cases.

The number of STS neurons per 50 μm-thick section in non-demented control subjects was stable across the sixth to ninth decades of life (average ± SD, $9.6 \pm 1.0 \times 10^4$). In AD, more than 50 % of the neurons were lost (average ± SD, $4.8 \pm 2.3 \times 10^4$; Fig. 4). Neuronal loss increased in parallel with the duration and severity of illness, beginning with almost no loss in the cases with shortest duration (< 2 years). In contrast with the marked loss of neurons in EC in patients with very mild dementia, preliminary analysis of the STS in the same CDR = 0.5

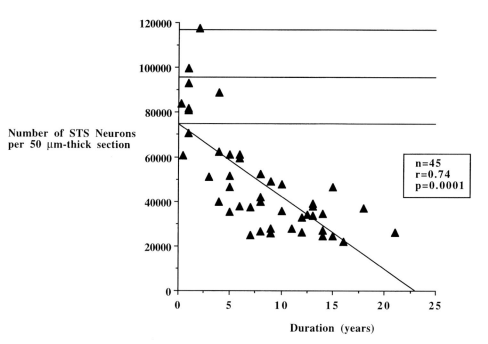

Fig. 4. The amount of neuronal loss in the STS in the AD group (n = 45) correlated with duration of illness (r = 0.74, p = 0.0001). Horizontal lines indicate the average number of neurons ± 2SD in the control group (n = 28).

patients in this study showed no difference in STS neuron number compared to non-demented controls. Thereafter there is a fairly rapid loss, reaching an apparent maximum of about 70 % in the cases with severest cognitive impairment or longest duration (15–20 years; Fig. 4). Interestingly, this is just about the same degree of loss as occurs, although at an earlier point in the disease, in the EC. These data suggest that there is a subpopulation of about one quarter to one third of neurons that are simply not vulnerable to AD neuropathological changes, even though they are within highly vulnerable cytoarchitectural regions.

Neuronal Loss Parallels but Outstrips NFT, and is Unrelated to SP in High Order Association Cortex

The assessment of the number of NFT in STS of AD brains shows that the amount of NFT increased also in close relation with the number of neurons lost and with the progression of dementia (Fig. 5). However, when absolute numbers of neurons and NFT are compared, neuronal loss in STS surprisingly outstrips by almost one order of magnitude the number of NFT. Taking into account the fact that the majority of NFT in the STS (more than 90 %) are intracellular, the ratio

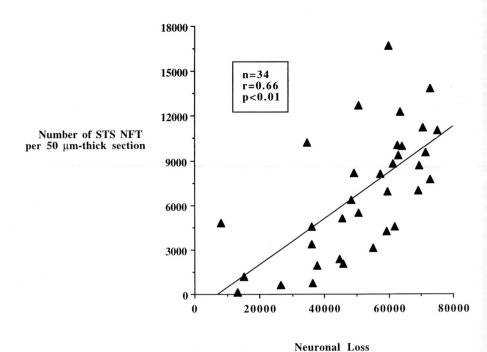

Fig. 5. The NFT number in the STS of AD brains (n = 34) correlated with the amount of neuronal loss (r = 0.66, p < 0.01). Neuronal loss was calculated by subtracting the number determined in each AD brain from the average of the control group.

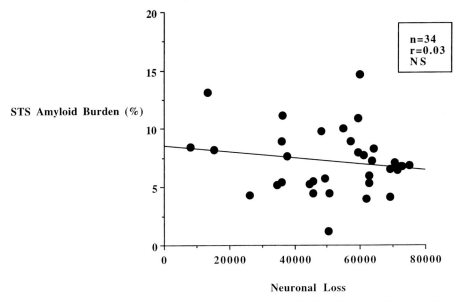

Fig. 6. The percentage of STS covered by Aβ (amyloid burden) in AD (n = 34) brains did not correlate with degree of neuronal loss (r = 0.03; NS).

of neuronal loss to extracellular or tombstone NFT probably approaches two orders of magnitude. Thus, only about 1–2 % of neuronal loss in the STS can be attributed to neurons for which an extracellular "tombstone" remains. This observation suggests one of two possibilities: 1) NFT might be formed, be involved in the death of neurons, and then be cleared leaving no trace, or 2) the majority of neuronal loss in advanced AD occurs through a non-NFT mechanism. The first possibility cannot be excluded, but the relatively insoluble nature of NFT qualifies them more as permanent marker of neuronal loss than as indices of a transient stage in cell death. The second possibility suggests that stereological techniques demonstrate an underappreciated alteration in AD brain that numerically is of substantially greater impact than NFT.

In contrast to the correlation between NFT development and neuronal loss, the number of SP and the percentage of the STS covered by Aβ (amyloid burden) are not related to neuronal loss, number of NFT or duration of disease in AD (Fig. 6). These data reinforce the previous conclusion that memory loss (usually impairment of delayed recall), which is often the earliest symptom in AD, stems from neuronal depopulation in the EC and other medial temporal lobe structures. During this early stage of dementia, even high order association cortices are relatively preserved in terms of neuronal numbers. As the clinical illness evolves from barely detectable memory loss to widespread cognitive failure with damage in visual and auditory processing systems, there is also a progressive loss of STS neurons (Fig. 7).

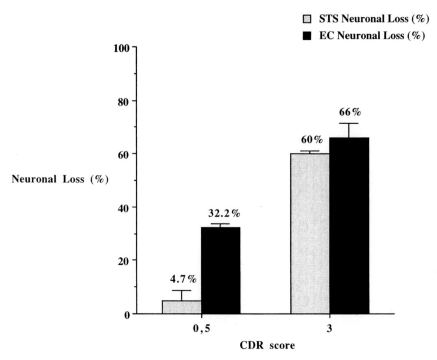

Fig. 7. Neuronal loss in the EC and STS at different stages of dementia (data shown are from Gómez-Isla et al. 1996). Recalculating the EC data for a single 50-micrometer thick section (technically comparable to the STS data) gives a loss in EC of 23 % in CDR 0.5 cases and about 60 % in CDR 3 cases.

Molecular epidemiology highlights the role of Aβ in Alzheimer's disease

By contrast to neurofibrillary tangles, which qualitatively and quantitatively match with clinical symptoms, Aβ deposition in the neuropil does not correlate with neuronal loss or with degree or duration of dementia. Despite this finding, compelling genetic evidence shows that Aβ is very likely a key player in AD pathophysiology (reviewed in Hyman and Tanzi 1995; Hyman 1996). Mutations in the amyloid precursor protein lead to autosomal dominant AD (Goate et al. 1991); inheritance of Down syndrome (Mann and Esiri 1989), apoE ε4 (Rebeck et al. 1993; Gómez-Isla et al. 1996) or presenilin mutations (Gomez-Isla, manuscript in preparation) lead to AD with increased Aβ deposition. Why is there a discrepancy between the lack of clinical-pathological correlates with Aβ and the genetic information implicating Aβ in dementia?

Since Aβ in plaques is an almost insoluble precipitate, it seems reasonable to assume that individual SP would get bigger as time went on, and that the amyloid burden would continue to grow. Surprisingly, our quantitative data refute both ideas. We noted earlier that the amount of amyloid deposited was not related to the duration or severity of dementia in AD cases that had been followed prospectively (Arriagada et al. 1992a; Hyman et al. 1993). By analyzing the size and shape of Aβ deposits in AD cases of varying duration of illness, we have also shown that the size of SP does not differ from early in the disease process to much later (Hyman et al. 1993, 1995). These results suggest that Aβ does not simply continue to accumulate in the AD brain, either within individual plaques or in large populations of plaques. We postulated that Aβ accumulates in a kinetically defined process up to a "steady state," in which continued deposition is offset by clearance mechanisms (Hyman et al. 1993, 1995). Despite the fact that Aβ is a rather insoluble compound, the plausibility of this argument is enhanced by recent data suggesting that, in vitro, microglia can ingest Aβ aggregates and even amyloid cores (Frautschy et al. 1992; Puresce et al. 1996).

Moreover, we have continued to develop the idea that amyloid deposits exist in steady state. Further detailed analysis of plaque shape distribution suggests that the size distribution fits a log-normal distribution, a size distribution characteristic of amphipathic substances in an aqueous environment (Hyman et al. 1995). In collaboration with Gene Stanley and colleagues at Boston University's Polymer Physics Department, we have generated a dynamic model that leads to experimentally verified predictions about plaque shape size and geometric properties. The major underlying assumption of this model is that there is a kinetic process that leads to clearance of Aβ. In fact, if one does not invoke such a mechanism, the model quickly "explodes;" all available space becomes occupied by amyloid. Both the log-normal distribution and the model predict a porous microarchitecture of individual senile plaques. Using Aβ immunostaining and laser scanning confocal microscopy, we have been able to verify this idea experimentally, showing that Aβ deposits are sponge-like in shape in their native configuration (in non-paraffin embedded material). Consistent with this microarchitecture, however, the neuropil "lawn" of synaptophysin immunoreactivity does not appear to be affected by many plaques, including diffuse deposits. However, compact plaques appear to occupy space in that the "lawn" of synapses usually spread throughout the neuropil is disrupted at the heart of a compact plaque, or synaptophysin immunoreactivity reveals abnormal swollen axonal patterns (Fig. 8) (Masliah et al. 1990, 1991, 1993a; Irizarry et al. 1996). Confocal microscopy studies suggest that there is extensive synapto-axonal damage with both degenerating and regenerative axonal elements contributing to a plaque (Masliah et al. 1993b). Masliah and colleagues have suggested that long cortico-cortical axons that terminate in the plaque are usually branched and distorted in their terminal site (Masliah et al. 1993b). These data are consistent with a model in which Aβ deposits form, may be associated with disruption of the neuropil and hence of synaptic structure, and then resolve, leaving behind (perhaps) neuropil threads and distributed pathological lesions in terms of synaptic alterations. If this model is accurate, the number of plaques observed at any time reflects a steady state rather than a measure of accumulated damage. Assuming for a moment that such a model of the life history of senile plaques is plausible, how does this help to address the underlying problem of how Aβ deposition is related to neuronal loss and ultimately to cognitive impairment?

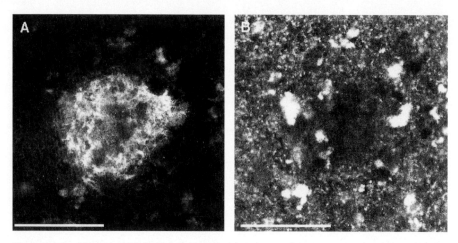

Fig. 8. Confocal image of (A) bodipy-fluorescein-labeled anti-Aβ (R1282, courtesy of Dr. D. Selkoe) counterstained with (B) Cy3-labeled anti-synuclein monoclonal H3C (courtesy of Dr. David Clayton). Synuclein, a protein found in synaptic specializations, is decreased in the core of many compact plaques (Irizarry et al. 1996).

Distributed pathological changes and neural system collapse

These data lead us to postulate that impairment of axon collaterals by distributed pathological changes may contribute to neural system collapse. We and others have suggested that disruption of projections between and among limbic and association areas leads to cognitive impairment. Could it also contribute to ongoing neurodegeneration? The consequences of disruption of synaptic contacts likely depends upon a variety of factors, including developmental state, degree of disruption, and percent of total contacts disrupted. In AD, there is strong evidence to suggest that a result of the entorhinal cortex lesions noted above is a plasticity event leading to reinnervation of the perforant pathway terminal zone (Geddes et al. 1985; Hyman et al. 1987). Moreover, anatomical fields downstream from the major projection neuron loss are metabolically impaired, consistent with a functional deafferentation (Simonian et al. 1994). Chronic EC lesion is associated with loss of neurons in CA3 of Ammons horn (Poduri et al. 1995), and destruction of the perforant pathway leads to progressive loss of entorhinal layer II neurons. Destruction of either afferent or efferent projections can lead to metabolic down-regulation, atrophy, and even neuronal death in certain circumstances (Jones 1990). Two well-studied examples of the latter, fimbria-fornix lesion leading to septal neuron loss and disruption of connections between the thalamus and cortex, both lead to the idea that the loss of connections must be nearly total to lead to neuronal loss. This kind of near total ablation could occur fairly easily in the entorhinal-dentate gyrus perforant pathway projection, because the vast majority of this major projection from layer II of entorhinal cortex terminates in a restricted field, the outer portion of the molecular layer of the dentate gyrus. Although experimental data are to a great extent still lacking, a restricted terminal field is probably the exception rather than the rule. Instead, it is likely that the vast majority of projection neurons targeted in AD have widespread local collateral projections and even possibly longer collaterals to multiple brain regions; they are similarly characterized by afferents from multiple sources. Thus a single lesion (for example a senile plaque) would be unlikely to disrupt more than a very small percentage of a neuron's contacts even if it were strategically placed. However, a population of widely distributed plaques might, by chance, disrupt collateral axons, making neurons more vulnerable to deafferenting lesions. This idea is consistent with observations of disruption of the terminal fields of phosphorylated neurofilament immunopositive axons in plaques (Masliah et al. 1993b).

We suggest a new role for Aβ deposits in AD: as a source of "distributed partial deafferentation" rather than as a directly neurotoxic deposit. Thus, a single Aβ deposit does not do much harm to its microenvironment or even to neurons in its vicinity. For example, it is not unusual by confocal microscopy to see a neuron within the borders of a senile plaque. Nonetheless, as sufficient numbers of SP accumulate throughout the limbic and association areas, at a certain critical point an individual neuron may face a crisis whereby many of its projections (at any given moment) are or have been disrupted by Aβ deposits. Thus a large number of senile plaques (of sufficient mass) may lead to a generalized deafferentation syndrome. As NFT also accumulate in many of these same neural systems, the cumulative effect may be a loss of both afferents and efferents that lead to metabolic compromise and ultimately neuronal death. Together, these factors lead to a state of progressive breakdown of critical neural systems subserving limbic and association area function, whose integrity is critical for normal memory, language, and judgment. Ultimately, rather than a single neuropathological lesion being responsible for the dementia of AD, the answer to the question of whether tangles, plaques, or synaptic or neural loss is responsible for dementia may lie in an appreciation of destruction of neural systems at multiple levels by focal neuronal pathological changes (NFT and neuropil threads) and more distributed lesions within the neuropil, including synaptic loss and SP. These ultimately may contribute to a final path of neuronal loss and the collapse of neural systems.

Acknowledgments

We thank the Ipsen Foundation for supporting this conference, NIH grants AG08487 and AG05134. We thank Drs Gary Van Hoesen, John Growdon, Mike Irizarry, H. Eugene Stanley, Sergei Buldyrev and their colleagues at Boston University Center for Polymer Physics, and Maggie Keane for helpful discussions. We thank Joel Price, Daniel McKeel, John Morris and their colleagues at Washington University ADRC, and Ronald Petersen, Joseph Parisi, and the Mayo Clinic Alzheimer Disease Center for collaborative study of clinical-pathological materials.

References

Arnold SE, Hyman BT, Flory J, Damasio AR, Van Hoesen GW (1991) The topographical and neuro-anatomical distribution of neurofibrillary tangles and neuritic plaques in cerebral cortex of patients with Alzheimer's disease. Cereb Cortex 1: 103–116

Arriagada PV, Growdon JH, Hedley-Whyte ET, Hyman BT (1992a) Neurofibrillary tangles but not senile plaques parallel duration and severity of Alzheimer disease. Neurology 42: 631–639

Arriagada PV, Marzloff K, Hyman BT (1992b) Distribution of Alzheimer-type pathological changes in nondemented elderly matches the pattern in Alzheimer's disease. Neurology 42: 1681–1688

Ball MJ (1978) Topographic distribution of neurofibrillary tangles and granulovacuolar degeneration in hippocampal cortex of aging and demented patients. A quantitative study. Acta Neuropathol 42: 73–80

Barnes CL, Pandya DN (1992) Efferent cortical connections of multimodal cortex of the superior temporal sulcus in the rhesus monkey. J Comp Neurol 318: 222-244

Berg L, McKeel DW, Miller JP, Baty J, Morris JC (1993) Neuropathological indexes of Alzheimer's disease in demented and nondemented persons aged 80 and older. Arch Neurol 50: 349–358

Blessed G, Tomlinson BE, Roth M (1968) The association between quantitative measures of dementia and of senile change in the cerebral grey matter of elderly subjects. Brit J Psychiatr 114: 797–811

Bouras C, Hof P, Giannakopoulos P, Michel JP, Morrison JH (1994) Regional distribution of neurofibrillary tangles and senile plaques in the cerebral cortex of elderly patients: a quantitative evaluation of a one-year autopsy population in a geriatric hospital. Cereb Cortex 4: 138–150

Bouras C, Hof PR, Morrison JH (1993) Neurofibrillary tangle densities in the hippocampal formation in a nondemented population define subgroups of patients with differential early pathological changes. Neurosci Lett 153: 131–135

Braak H, Braak E (1991) Neuropathological staging of Alzheimer related changes. Acta Neuropathol 82: 239–259

Cabalka LM, Hyman BT, Goodlett CR, Ritchie TC, Van Hoesen GW (1992) Alteration in the pattern of nerve terminal protein immunoreactivity in the perforant pathway in Alzheimer's disease and in rats after entorhinal lesions. Neurobiol Aging 13: 283–291

Coggeshall RE (1992) A consideration of neural counting methods. Trends Neurosci 15: 9–13

De Kosky ST, Scheff SW (1990) Synapse loss in frontal cortex biopsies in Alzheimer's disease: Correlation with cognitive severity. Ann Neurol 27: 457–464

Frautschy S, Cole G, Baird A (1992) Phagocytosis and deposition of vascular beta amyloid in rat brains injected with Alzheimer beta amyloid. Am J Pathol 140: 1389–1399

Games D, Adams D, Alessandri R, Barbour R, Berthelette P, Blackwell C, Carr T, Clemens J. Donaldson T, Gillespie R, Guido T, Hagopian S, Johnson-Wood K, Khan I, Lee M, Leibowitz P, Lieberburg I, Little S, Masliah E, McConlogue L, Montoya-Azvala M, Mucke L, Paganini L, Penniman E, Power M, Schenk D, Seubert P, Snyder B, Soriano F, Tan H, Vitale J, Wadsworth S, Wolozin B, Zhao J (1995) Alzheimer-type neuropathology in transgenic mice overexpressing V717F beta-amyloid precursor protein. Nature 373: 523–527

Geddes JW, Monaghan DT, Cotman CW, Lott IT, Kim RC, Chui HC (1985) Plasticity of hippocampal circuitry in Alzheimer's disease. Science 230: 1179–1181

Goate A, Chartier-Harlin M-C, Mullan M, Brown J, Crawford F, Fidani L, Giuffra L, Haynes A, Irving N, James L, Mant R, Newton P, Rooke K, Roques P, Talbot C, Pericak-Vance M, Roses A, Williamson R, Rossor M, Owen M, Hardy J (1991) Segregation of a missense mutation in the amyloid precursor protein gene with familial Alzheimer's disease. Nature 349: 704–707

Gómez-Isla T, Price J, McKeel D, Morris J, Growdon J, Hyman B (1996a) Profound loss of layer II entorhinal cortex neurons occurs in very mild Alzheimer's disease. J Neurosci 16: 4491–4500.

Gómez-Isla T, West HL, Rebeck GW, Harr SD, Growdon JH, Locasio JT, Perls TT, Lipsitz LA, Hyman BT (1996b) Clinical and pathological correlates of apolipoprotein E ε4 in Alzheimer's disease. Ann Neurol 39: 62–70

Hof PR, Bierer LM, Perl DP, Delacourte A, Buee L, Bouras C, Morrison J (1992) Evidence of early vulnerability of the medial and inferior aspects of the temporal lobe in an 82 year old patient with preclinical signs of dementia. Arch Neurol 49: 946–953

Hof PR, Cox K, Morrison JH (1990) Quantitative analysis of a vulnerable subset of pyramidal neurons in Alzheimer's disease: I. Superior frontal and inferior temporal cortex. J Comp Neurol 301: 44–54

Hyman B (1996) Alzheimer's disease or Alzheimer's diseases? Clues from molecular epidemiology. Ann Neurol 40: 135–136

Hyman B, Tanzi R (1995) Molecular epidemiology of Alzheimer disease. N Engl J Med 333: 1283–1285

Hyman BT, Gómez-Isla T (1994) Alzheimer's disease is a laminar, regional, and neural system specific disease. Neurobiol Aging 15: 353–354

Hyman BT, Kromer LJ, Van Hoesen GW (1987) Reinnervation of the hippocampal perforant pathway zone in Alzheimer's disease. Ann Neurol 21: 259–267

Hyman BT, Marzloff K, Arriagada PV (1993) The lack of accumulation of senile plaques or amyloid burden in Alzheimer's disease suggests a dynamic balance between amyloid deposition and resolution. J Neuropathol Exp Neurol 52: 594–600

Hyman BT, Van Hoesen GW, Damasio AR (1990) Memory-related neural systems in Alzheimer's disease: An anatomical study. Neurology 40: 1721–1730

Hyman BT, Van Hoesen GW, Damasio AR, Barnes CL (1984) Alzheimer's disease: Cell specific pathology isolates the hippocampal formation in Alzheimer's disease. Science 225: 1168–1170

Hyman BT, Van Hoesen GW, Kromer LJ, Damasio AR (1986) Perforant pathway changes and the memory impairment of Alzheimer's disease. Ann Neurol 20: 473–482

Hyman BT, West HL, Rebeck GW, Buldyrev SV, Mantegna RN, Ukleja M, Havlin S, Stanley HE (1995) Quantitative analysis of senile plaques in Alzheimer disease: observation of log-normal size distribution and molecular epidemiology of differences associated with apolipoprotein E genotype and trisomy 21 (Down syndrome). Proc Natl Acad Sci USA 92: 3586–3590

Irizarry M, Kim T-W, McNamara M, Tanzi R, George J, Clayton D, Hyman B (1996) Characterization of the precursor protein of the non-Aβ component of senile plaques (NACP) in the human central nervous system. J Neuropathol Exp Neurol, in press.

Jones EG (1990) The role of afferent activity in the maintenance of primate neocortical function. J Exp Biol 153: 155–176

Khachaturian ZS (1985) Diagnosis of Alzheimer's disease. Arch Neurol 42: 1097–1105

Lewis D, Campbell M, Terry R, Morrison J (1987) Laminar and regional distributions of neurofibrillary tangles and neuritic plaques in Alzheimer's disease: a quantitative study of visual and auditory cortices. J Neuroscience 7: 1799–808

Mann DMA, Esiri MM (1989) The pattern of acquisition of plaques and tangles in the brains of patients under 50 years of age with Down syndrome. J Neurol Sci 89: 169–179

Masliah E, Mallory M, Deerinck T, DeTeresa R, Lamont S, Miller A, Terry R, Carragher B, Ellisman M (1993a) Re-evaluation of the structural organization of neuritic plaques in Alzheimer's disease. J Neuropathol Exp Neurol 52: 619–632

Masliah E, Mallory M, Hansen L, Alford M, DeTeresa R, Terry R (1993b) An antibody against phosphorylated neurofilaments identifies a subset of damaged association axons in Alzheimer's disease. Am J Pathol 1993: 871–882

Masliah E, Terry R, Alford M, DeTeresa R, Hansen L (1991) Cortical and subcortical patterns of synaptohysin-like immunoreactivity in Alzheimer disease. Am J Pathol 138: 235–246

Masliah E, Terry RD, Mallory M, Alford M, Hansen LA (1990) Diffuse plaques do not accentuate synapse loss in Alzheimer's disease. Am J Pathol 137: 1293–1297

Mirra SS, Heyman A, McKeel D (1991) The consortium to establish a registry for Alzheimer's disease (CERAD) Part II. Standardization of the neuropathologic assessment of Alzheimer's disease. Neurology 41: 479–486

Nagy Z, Esiri M, Jobst K, Morris J, King E-F, McDonald B, Litchfield S, Smith A, Barnetson L, Smith A (1995) Relative roles of plaques and tangles in the dementia of Alzheimer's disease: Correlations using three sets of neuropathological criteria. Dementia 6: 21–31

Pandya DN, Yeterian EH (1985) Architecture and connections of cortical association areas. New York, Plenum Press.

Poduri A, Beason-Held L, Moss M, Rosene D, Hyman B (1995) CA3 neuronal degeneration follows chronic entorhinal cortex lesions. Neurosci Lett 197: 1–4

Price D, Davis P, Morris J, White D (1991) The distribution of tangles, plaques and related immuno-histochemical markers in healthy aging and Alzheimer's disease. Neurobiol Aging 12: 295–312

Puresce D, Ghosh R, Maxfield F (1996) Microglial cells bind and internalize aggregates of the Alzhei-mer's disease beta-amyloid peptide by the type I scavenger receptor. Mol Biol Cell 6: 94a

Rebeck GW, Reiter JS, Strickland DK, Hyman BT (1993) Apolipoprotein E in sporadic Alzheimer's dis-ease: Allelic variation and receptor interactions. Neuron 11: 575–580

Simonian N, Rebeck G, Hyman B (1994) Functional integrity of neural systems related to memory in Alzheimer disease. Prog Brain Res 100: 245–254

Terry RD, Masliah E, Salmon DP (1991) Physical basis of cognitive alterations in Alzheimer's disease: Synapse loss is the major correlate of cognitive impairment. Ann Neurol 41: 572–580

Terry RD, Peck A, DeTeresa R, Schechter R, Horoupian DS (1981) Some morphometric aspects of the brain in senile dementia of the Alzheimer type. Ann Neurol 10: 184–192

Van Hoesen GW (1993) The modern concept of association cortex. Curr Opinion Neurobiol 3: 150–154

Van Hoesen GW, Solodkin A (1993) Some modular features of temporal cortex in humans as revealed by pathological changes in Alzheimer's disease. Cerebral Cortex 3: 465–475

West M (1993) New stereological methods for counting neurons. Neurobiol Aging 14: 275–285

West M, Coleman P, Flood D, JC T (1994) Differences in pattern of hippocampal neuronal loss in nor-mal ageing and Alzheimer's disease. Lancet 344: 769–72

Distributed Hierarchical Neural Systems for Visual Memory in Human Cortex

J. V. Haxby, *V. P. Clark, M. Courtney*

Summary

The visual cortex in human and nonhuman primates consists of multiple areas that are hierarchically organized into processing pathways. A ventral pathway in the occipitotemporal cortex is critical for the perception of object identity, and a dorsal pathway in the occipitoparietal cortex is critical for the perception of the spatial relations among objects, the perception of movement, and the direction of movements toward objects. The ventral object vision and dorsal spatial vision pathways have projections into different parts of the prefrontal cortex. Functional brain imaging studies of visual working memory for faces and spatial locations reveal activity throughout the ventral object vision and dorsal spatial vision pathways, respectively. Different prefrontal areas are associated with working memory for faces and for locations. Face working memory is selectively associated with regions in the inferior and mid-frontal cortex. Location working memory is selectively associated with a more dorsal and posterior region in the superior frontal sulcus. These prefrontal regions demonstrate sustained activity during working memory delays demonstrating the role these areas play in maintenance of an active representation of a visual working memory. Functional brain imaging studies of long-term episodic memory also demonstrate a mnemonic role for prefrontal areas. Encoding new long-term memories for faces was associated with activity in the right hippocampus and in the left prefrontal and inferior temporal cortex. Recognition of memorized faces, on the other hand, was not associated with hippocampal activity but was associated with increased activity in the right prefrontal and parietal cortex. This research shows that widely distributed neural systems are associated with working and episodic visual memory. Mnemonic functions involve the concerted activity of multiple regions in the posterior extrastriate and prefrontal cortices.

---

* Section on Functional Brain Imaging, NIMH, Building 10, Room 4C110, 10 Center Drive, MSC 1366, Bethesda, MD 20892-1366, USA

B. T. Hyman / C. Duyckaerts / Y. Christen (Eds.)
Connections, Cognition, and Alzheimer's Disease
© Springer-Verlag Berlin Heidelberg 1997

Introduction

Information processing in the human brain involves the concerted activity of multiple, spatially distributed cortical regions. Major goals of cognitive neuroscience include identifying the regions that participate in different information processing operations, characterizing the functional role played by each region, and modeling the interactions between regions. The ensemble of regions that participate in a specific perceptual or cognitive function will be referred to here as a functional neural system. This functional definition, as compared to an anatomical definition, bases the identification of regions that comprise a functional neural system on their coactivation and functional interactions during the performance of a perceptual or cognitive function. The concept of a functional neural system is proposed here to guide discussion of the neural substrate for human visual perception and human memory. In particular, functional neural systems are proposed as a principle of organization that better captures the contribution that functional brain imaging has made to our understanding of the neural processes underlying visual perception and memory.

Functional brain imaging measures local hemodynamic changes, and these changes are used as indices of changes in neural activity. In contrast to most methods for studying neural function that examine one neuron or one region at a time, functional brain imaging affords measures of simultaneous activity in all regions of the brain, making it ideally suited for investigating the concerted activity in distributed functional neural systems. Vision and visual memory have proven to be particularly fruitful domains for functional brain imaging studies of human neural systems. Vision and visual memory are functions that we share with nonhuman primates, providing an animal model that allows more invasive, experimental research. An extensive literature on visual neuroanatomy and neurophysiology in the monkeys provides a critical foundation for human vision research.

Visual Processing Pathways in the Macaque

The visual cortex in the macaque monkey occupies over half of the cortical surface (Desimone and Ungerleider 1989; Felleman and Van Essen 1991). This cortex is comprised of over 30 areas. These areas are defined by their retinotopic organization, functional properties, patterns of connections with other areas, myelo- and cyto-architecture, and the effects of lesions. They are organized hierarchically into at least two processing pathways (Ungerleider and Haxby 1994; Ungerleider and Mishkin 1982). Both pathways originate in primary visual cortex, area V1, and have relays in early extrastriate areas V2 and V3. Further projections to areas in the ventral occipitotemporal cortex comprise a pathway that is critical for perception of object identity. A second pathway comprised of areas in dorsal occipitoparietal cortex is critical for perception of the spatial relations among objects, for perception of motion, and for the visual guidance of movements (Fig. 1).

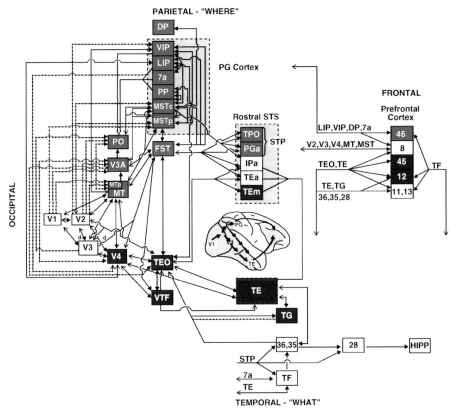

Fig. 1. The organization of visual cortex in the macaque monkey. Regions in the dorsal spatial vision pathway are shown in green, and regions in the ventral object vision pathway are shown in red. Lines between regions indicate known anatomical connections. Solid lines represent connections arising from both central and peripheral (from Ungerleider 1995)

The multiple areas in the ventral object vision and dorsal spatial vision pathways display properties that suggest a hierarchical organization. Feedforward connections from early to later areas typically originate in layer III and terminate in layer IV, whereas feedback connections from later to earlier areas originate in layer V and terminate in layers above and below layer IV, indicating a dominant direction for the flow of information with extensive feedback (Felleman and Van Essen 1991). The stimulus attributes to which cells respond become more complex and abstract in later visual areas (Desimone and Ungerleider 1989). For example, in the object vision stream, cells in early visual areas respond selectively to simple features such as oriented edges in restricted parts of the visual cells. Cells in later visual areas in inferior temporal cortex respond selectively to complex patterns and, in some cases, faces. The receptive fields of these cells are much larger, indicating they respond to complex patterns or objects independent of retinal location.

The ventral object vision and dorsal spatial vision pathways project to different regions in the prefrontal cortex (see Fig. 1). These prefrontal regions play a critical role in another high order cognitive function, namely working memory (Fuster 1990; Goldman-Rakic 1990). In delayed match-to-sample, delayed alternation, and delayed response tasks, some cells in the prefrontal cortex demonstrate sustained activity over the delay period. This sustained activity is believed to play a role in maintaining an active representation of visual information in working memory. The prefrontal extensions of the spatial and object vision pathways appear to be specialized for their roles in spatial and object working memory, respectively (Wilson et al. 1993). Sustained activity during memory delays is also seen in inferior temporal neurons, but, unlike in prefrontal cells, this activity is disrupted by intervening, distracting stimuli. Thus, the maintenance of an active, working memory representation of a stimulus after that stimulus has been removed from sight appears to involve the participation of very late regions in the visual processing pathways, with a dominant role played by prefrontal areas and some support from inferior temporal areas.

Two Visual Processing Pathways in Human Extrastriate Cortex

Functional brain imaging studies of the organization of the human visual cortex demonstrate extensive homologies with nonhuman primates. Mapping of early visual areas based on their retinotopic organization using functional magnetic resonance imaging (fMRI) has demonstrated that these areas (V1, V2, V3, VP, V4) have a remarkably similar spatial organization in humans and nonhuman primates (Sereno et al 1995). Studies of motion perception have identified a region in the human brain that is the homologue of area MT/V5 in the monkey (Corbetta et al. 1991; Tootell et al. 1995; Watson et al. 1993; Zeki et al. 1991).

We have investigated whether later extrastriate regions demonstrate organization into ventral object vision and dorsal spatial vision streams (Clark et al. 1996; Haxby et al. 1991, 1994a). In these studies, functional brain imaging was used to measure local hemodynamic changes that are indices of changes in neural activity. We have used two methods of functional brain imaging: positron emission tomography (PET) and fMRI. Using bolus injections of water labeled with the positron emitting isotope oxygen-15, PET affords measures of regional cerebral blood flow (rCBF). T2*-weighted MRI scans use an endogenous contrast agent, hemoglobin, to detect changes in blood oxygenation. Blood oxygenation level-dependent (BOLD) fMRI demonstrates changes in intensity related to the paramagnetic property of deoxyhemoglobin. BOLD fMRI works because the increase in rCBF induced by a change in neural activity is greater than the increase in oxygen metabolism, resulting in higher blood oxygenation and increased T2*-weighted MRI signal.

In a PET-rCBF study of visual perception, activation of ventral object vision and dorsal spatial vision pathways was induced by performance of match-to-sample tests in which the match was based on face identity or on spatial location

(Haxby et al. 1994b). Identical stimuli were used for these two tasks. Regions that were activated by these tasks were identified by comparing rCBF during face matching or location matching to rCBF during the performance of a sensorimotor control task. Early visual areas in occipital cortex were activated by both tasks. The face matching but not the location matching task also activated the ventral occipitotemporal cortex, centered on the fusiform gyrus. By contrast, the location matching but not the face matching task also activated the dorsal occipital, superior parietal, and intraparietal sulcal cortex. These results confirmed two hypotheses. First, human extrastriate cortex did demonstrate an overall organization into ventral and dorsal pathways specialized for object and spatial vision, respectively. Second, human visual cortex demonstrated hierarchical organization insofar as early areas show less functional specialization as compared to later areas.

A functional MRI study using the same tasks has replicated the results of the PET-rCBF study of spatial and object vision (Fig. 2, Fig. 3). With fMRI the regions of activation in each individual subject can be precisely delineated, as compared to PET-rCBF studies in which the results for a group of subjects is averaged to increase signal-to-noise ratios and sensitivity. The cortical areas activated by the face matching task were found to have discrete borders, were of smaller spatial extent than indicated by the group average PET-rCBF results, and demonstrated small variations in their precise anatomical location relative to sulcal landmarks (Clark et al. 1996).

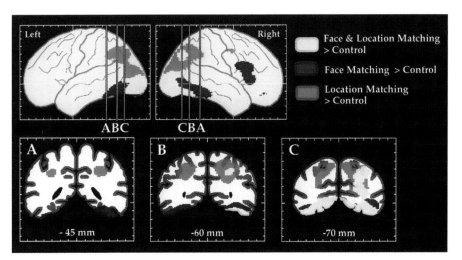

Fig. 2. The dorsal spatial vision and ventral object pathways in humans as demonstrated in a PET-rCBF study of face and location perception. Areas shown in green had significantly increased rCBF during a location matching but not during a face matching task, as compared to rCBF during a sensorimotor control task. Areas shown in red had significantly increased rCBF during the face matching but not the location matching task. Areas shown in yellow had significantly increased rCBF during both face and location matching. Maximum intensity projections are shown on the lateral views of the left and right hemisphere. Coronal sections are shown for locations indicated on the lateral views (adapted from Haxby et al. 1994).

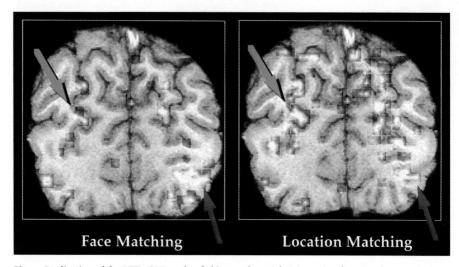

Face Matching   **Location Matching**

Fig. 3. Replication of the PET-rCBF study of object and spatial vision using functional magnetic resonance imaging. Results from a single subject are shown. These coronal sections are 3 cm from the occipital pole. Note the ventral lateral occipital areas that are more activated by face matching than by location matching and the dorsal occipital areas that were more activated by location matching than by face matching.

The locations of the object and spatial vision pathways in the human cortex is similar to those in the macaque brain, but with some important differences. The object vision pathway is located more exclusively on the ventral surface of the temporal lobe in the human as compared to the macaque and does not extend as far anteriorly. The spatial vision pathway has a more superior parietal location. These differences suggest that these phylogenetically older visual areas were displaced in human brain evolution. Displacement away from the posterior perisylvian cortex may be related to the development of language. Displacement away from the temporal pole may be related to the role this cortex may play in semantic and lexical knowledge about objects (Damasio et al. 1996; Nobre et al. 1994).

The object and spatial vision pathways revealed by this study were comprised of multiple areas, supporting the hypothesis that they are organized as pathways. Both pathways also included regions in the right frontal cortex. The ventral pathway included a region, in the right inferior frontal cortex. The dorsal pathway included a region in the superior frontal cortex, with a deep location that indicates it is in the superior frontal sulcus. In a lateral view, this region looks like it is centered in the premotor cortex. Our results and those of others (Corbetta et al. 1993; Jonides et al. 1993), however, suggest that activity in this region is not dependent on motor planning or motor preparedness, suggesting that its functional properties are more characteristic of prefrontal cortex.

Human Neural Systems for Visual Working Memory

Finding prefrontal areas associated with the ventral object vision and dorsal spatial vision pathways led us to ask whether these areas played a dominant role in maintaining an active working memory representation of object or spatial visual information after relevant stimuli are removed from view.

In a PET-rCBF study of visual working memory, we used tasks that used the same stimuli for object and spatial working memory (Courtney et al. 1996). In each item the subject tried to remember the identity or location of three faces presented sequentially. Immediately after the presentation of the third face, a test stimulus, namely a face in a location, was shown. For the face working memory task, the subject indicated whether that face was in the memory set, regardless of the location in which it was initially presented. For the location working memory task, the subject indicated whether the location of the test stimulus was one of the locations in the memory set, regardless of the identities of the faces used to mark those locations.

Comparison of rCBF during the performance of these two tasks again revealed differences in the ventral object vision and dorsal spatial vision pathways (Fig. 3). In the ventral pathway, rCBF differences were found in a right hippocampal region and in the right thalamus, regions that were not differentially activated in the visual perception study. In the prefrontal cortex, large activations in right inferior and mid frontal areas were associated with the face working memory task. The spatial working memory task was associated with bilateral activations in the superior frontal sulcus. Because both of these tasks involved identical motor responses, this activation cannot be attributed to differences in motor planning or motor preparedness (Fig. 4).

The size of the prefrontal activations in this study suggests that they play a dominant role in working memory function, but activity related to perception could not be clearly distinguished from activity related to working memory. To test whether the posterior extrastriate and prefrontal areas could be distinguished by the extent to which they participate differentially in perception and working memory, respectively, a parametric PET-rCBF study of face working memory was performed (Haxby et al. 1995). In this study, the delay between the presentation of a face to hold in working memory and the test stimulus was systematically varied. This variation was predicted to have differential effects on activity in areas with primarily perceptual function as compared to areas with working memory function. Duty cycle, that is the proportion of time devoted to one type of activity, varied differently for perception and active working memory. With shorter delays, a larger proportion of time was devoted to viewing pictures of faces. With longer delays, a smaller proportion of time was devoted to viewing faces and a larger proportion was devoted to holding a face in memory while viewing blank stimuli. It was predicted that areas with primarily perceptual function would demonstrate progressively smaller rCBF increases with longer memory delays. It was further predicted that areas with working memory function would show relatively constant levels of rCBF increase across delays. In the

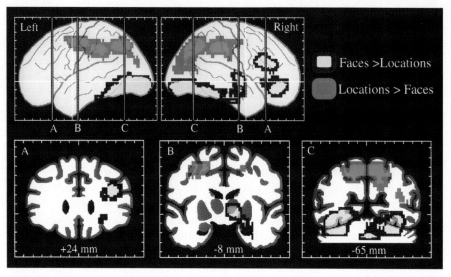

Fig. 4. Selective activation of the dorsal spatial vision and ventral object vision pathways and dorsal and vental prefrontal areas during spatial and face working memory tasks. Identical stimuli were used for the two working memory tasks. Areas shown in blue and green had significantly higher rCBF during the spatial, as compared to the face, working memory task. Areas shown in yellow and red had significantly higher rCBF during the face, as compared to the spatial, working memory task (adapted from Courtney et al. 1995).

macaque, these areas are known to contain neurons that respond both to stimulus presentation and during memory delays, and should, therefore, be active throughout the face working memory task, regardless of delay length.

Areas demonstrating a significant negative correlation between rCBF and memory delay were restricted to the ventral occipitotemporal cortex, in areas that were essentially identical to those identified by studies of face perception. This result indicated that these areas are active primarily during the presentation of stimuli, not during memory delays, and, therefore, play mostly a perceptual role in face working memory. By contrast, areas that were activated in the right and left prefrontal cortex demonstrated more sustained rCBF increases across all delays, indicating that they are active during memory delays as well as during stimulus presentation and, therefore, play an important role in the maintenance of a working memory representation. Moreover, the relationship between rCBF and retention interval differed significantly by hemisphere. Whereas right frontal rCBF showed a nonsignificant tendency to diminish at longer delays, left frontal cortex demonstrated the largest rCBF increases at the longest delays. These results indicate that, right frontal and left frontal areas both participate in face working memory, but that left hemisphere areas may be associated with a more durable working memory representation or strategy that subjects rely on increasingly with longer retention intervals. One possible explanation for this hemispheric difference is that left hemisphere activity is associated with a face repre-

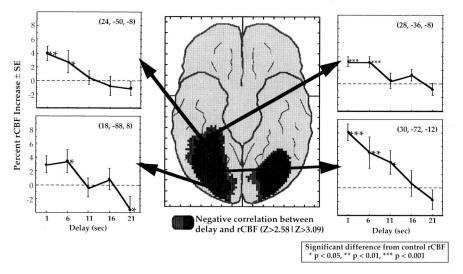

Fig. 5. Areas in the ventral extrastriate cortex that demonstrated significant negative correlations between memory delay length and rCBF during a parametric PET-rCBF study of face working memory. Delays varied from 1 to 21 seconds. Negative correlations indicated that these areas were active primarily during stimulus presentation (adapted from Haxby et al. 1995).

sentation that embodies the results of more analysis and elaboration, whereas right frontal activity is associated with a simpler, icon-like image of a face that is harder to maintain in working memory (Fig. 5, Fig. 6).

Long-Term Memory Encoding and Recognition of Faces

The time scale of working memory tasks makes it difficult to distinguish with PET-rCBF studies between activity related to perception and that related to memory and between activity related to the formation or encoding of a new memory and that related to the retrieval of that memory. To address this second functional distinction, we conducted a PET-rCBF study of long-term memory for faces. We obtained separate measures of rCBF while subjects encoded new long-term memories for faces and while they retrieved those memories in a face recognition task. Face encoding rCBF and face recognition rCBF were both compared to measures of rCBF that were obtained while subjects performed a face perception task that required no long-term memory function. As compared to a sensorimotor control task, face perception, encoding, and recognition all activated the ventral occipitotemporal areas that are associated with face perception. The areas that demonstrated additional activity during face encoding and during face recognition had almost no overlap. Face encoding, but not recognition, additionally activated the right hippocampus and areas in left prefrontal and left inferior temporal cortex. Face recognition, but not encoding, additionally acti-

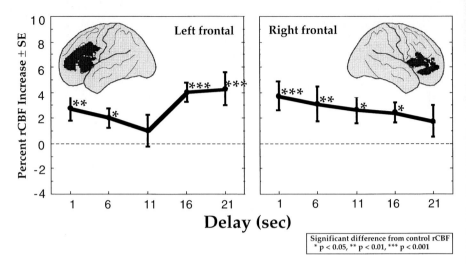

Fig. 6. Areas in right and left prefrontal cortex that were activated during performance of a face working memory task. Activity in these regions was significantly more sustained across memory delay lengths than was activity in the posterior ventral extrastriate cortex (Fig. 5), indicating that they are more active during the memory delays after the stimuli are removed from view. Note that whereas the right prefrontal region shows a tendency to have diminished rCBF at longer delays, the left prefrontal region shows the greatest rCBF increase at that delay (adapted from Haxby et al. 1995).

vated areas in the right prefrontal, bilateral inferior parietal, and bilateral inferior occipital cortex (Fig. 7).

These results demonstrate that the formation and retrieval of long-term memories for information processes by the ventral object vision pathway are realized by coactivation of that neural system and neural systems that have more general long-term memory functions.

It has long been known, based on studies of amnesia in humans and of models of human amnesia in nonhuman primates, that the hippocampus plays a critical role in long-term memory (Scoville and Milner 1957; Squire 1992). The results of our face memory study suggest that this structure plays a greater role at the time new memories are encoded as compared to when they are retrieved. The activation of the hippocampus that was seen in both face working memory studies, therefore, is probably attibutable to operations performed during the encoding phases of those tasks.

A large body of imaging research on both verbal and nonverbal long-term memory has consistently demonstrated the participation of the prefrontal cortical areas, with the left prefrontal cortex associated with memory encoding and the right prefrontal cortex associated with memory retrieval (Tulving et al. 1994). These studies show involvement of these regions in both verbal and nonverbal memory, with both visual and auditory presentation of stimuli, indicating that these systems have more general memory functions and are not part of the visual processing pathways.

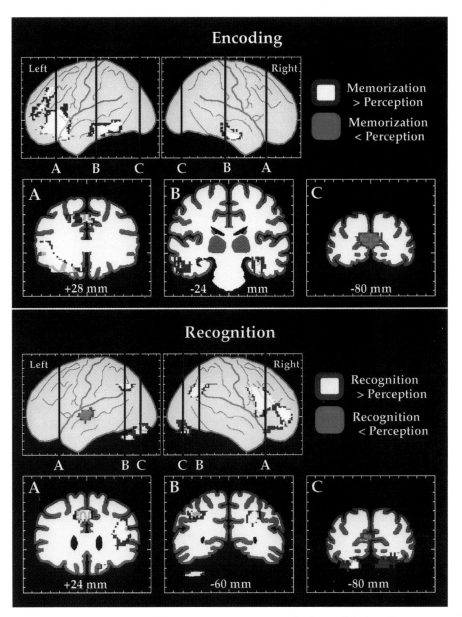

Fig. 7. Areas selectively activated during encoding new memories for faces and during subsequent recognition of those faces, as compared to a face perception baseline task (from Haxby et al. 1996).

Neural Systems

With functional brain imaging, the working of entire neural systems can be examined. Using these methods, we have identified multiple regions in an object vision pathway in the ventral occipitotemporal cortex and a spatial vision pathway in the dorsal occipitoparietal cortex. Both pathways demonstrate a hierarchical functional organization. Later areas are more specialized for processing object or spatial visual information.

Working memory involves an interaction between the extrastriate visual pathways and prefrontal areas. Activity in ventral extrastriate areas is predominantly associated with stimulus presentation whereas activity in prefrontal areas is associated with the maintenance of an active representation during memory delays. Different prefrontal areas are associated with spatial and face working memory. In face working memory, different prefrontal areas are recruited for retention intervals of different durations. It is not clear yet whether these prefrontal areas are the projection sites of the extrastriate pathways and specifially visual, or if they are also associated with working memory for other types of information (Fiez et al. 1996) or with supramodal working memory functions such as the central executive (Baddeley 1986; D'Esposito et al. 1995).

Similarly, long-term episodic memory for faces is associated with an interaction between ventral extrastriate visual cortex and other areas. The other areas appear to play a general role in episodic memory that is not specific to visual memory. The brain structures that subserve episodic memory encoding and retrieval demonstrate almost no overlap. Structures in the medial temporal lobe, long known to be critical for episodic memory, were active during encoding but not retrieval. Left prefrontal areas were associated with encoding whereas right prefrontal areas were associated with retrieval. It is not known whether these areas are the same as areas that are activated during working memory tasks. Performance of a working memory task by a normal individual may involve recruitment of episodic memory to provide a more durable, reserve representation that could be used if an active representation is disrupted. Conversely, prefrontal activations associated with memory encoding and retrieval may be due to encoding and retrieving the circumstances of the event during which learning occurred. Encoding and retrieval of the temporal-spatial context of an event is the hallmark of episodic memory, and working memory may maintain active representations of the information that comprises that context.

By affording simultaneous measures of activity in the entire brain, functional brain imaging draws our attention to neural activity organized at the level of systems. The performance of a cognitive act involves the activity of an entire system. Regions in these systems make different functional contributions to a cognitive act, and this contribution may or may not be critical for successful performance. Before the development of functional brain imaging, our understanding of the functional organization of the human brain was based primarily on neuropsychological studies to the effects of lesions and on assumed homologies with nonhuman primates. The functional organization of the nonhuman primate was, in turn, revealed by research methods that also focused attention on the function of single regions, namely single unit recording, controlled lesions, and tracing the connections from single regions. These research methods divert attention from the concerted activity in distributed neural systems and focus on defining the function of single regions. With these methods it is difficult to investigate how the function of an individual region may be altered by interactions with other regions, how regions working in concert may perform an operation that is difficult to discern from the function of single regions, and how the execution of a complete cognitive act involves the coordinated participation of multiple regions, each contributing to different components of that act. The function of individual regions may be best understood and modeled in the context of the neural systems in which they participate.

Functional brain imaging, by affording simultaneous measures of activity in all regions of the brain, can redirect the attention of cognitive neuroscience researchers to viewing functional neural systems as a fundamental principal of organization of function in the brain. Simultaneous measures of activity also allow analysis of the covariance the fluctuations in activity in different regions. Analytic methods for investigating inter-regional covariance as an indicator of functional connections promise to enhance further the usefulness of functional brain imaging for studying neural systems (Horwitz et al. 1992; McIntosh et al. 1994).

The two visual pathway hypothesis was based on a synthesis of information gathered from scores of studies using single unit recording, controlled lesions, and connection tracing. Functional brain imaging made it possible, for the first time, to observe the simultaneous functioning of the complete pathways and to observe interactions between these pathways and systems involved in memory processing.

References

Baddeley A (1986) Working memory. Oxford Univ Press New York

Clark VP, Keil K, Maisog JM, Courtney SM, Ungerleider LG, Haxby JV (1996) Functional magnetic resonance imaging (fMRI) of human visual cortex during face matching: A comparison with positron emission tomography (PET). NeuroImage, in press

Corbetta M, Miezin FM, Dobmeyer S, Shulman GL, Petersen SE (1991) Selective and divided attention during visual discriminations of shape, color, and speed: functional anatomy by positron emission tomography. J Neurosci 11: 2383–2402

Corbetta M, Miezin FM, Shulman GL, Petersen SE (1993) A PET study of visuospatial attention. J Neurosci 13: 1202–1226

Courtney SM, Ungerleider LG, Keil K, Haxby JV (1996) Object and spatial visual working memory activate separate neural systems in human cortex. Cereb Cortex 6: 39–49

Damasio H, Grabowski TJ, Tranel D, Hichwa RD, Damasio AR (1996) A neural basis for lexical retrieval. Nature 380: 499–505

Desimone R, Ungerleider LG (1989) Neural mechanisms of visual processing in monkeys. In: Goodglass H, Damasio AR (eds) Handbook of neuropsychology. Elsevier, Amsterdam, pp 267–300

D'Esposito M, Detre JA, Alsop DC, Shin RK, Atlas S, Grossman M (1995) The neural basis of the central executive system of working memory. Nature 378: 279–281

Felleman DJ, Van Essen DC (1991) Distributed hierarchical processing in the primate cerebral cortex. Cereb Cortex 1: 1–47

Fiez JA, Raife EA, Balota DA, Schwarz JP, Raichle ME, Petersen SE (1996) A positron emission tomography study of the short-term maintenance of verbal information. J Neurosci 16: 808–822

Fuster JM (1990) Behavioral electrophysiology of the prefrontal cortex of the primate. In: Uylings HMB, Van Eden JPC, De Bruin MA, Corner MA, Feenstra MGP (eds) Progress in brain research. Elsevier, Amsterdam, pp 313–323

Goldman Rakic PS (1990) Cellular and circuit basis of working memory in prefrontal cortex of nonhuman primates. In: Uylings HBM, Eden CGV, Bruin JPCD, Corner MA, Feenstra MGP (eds) Progress in brain research. Elsevier, Amsterdam, pp 325–336

Haxby JV, Grady CL, Horwitz B, Ungerleider LG, Mishkin M, Carson RE, Herscovitch P, Schapiro MB, Rapoport SI (1991) Dissociation of spatial and object visual processing pathways in human extrastriate cortex. Proc Natl Acad Sci USA 88: 1621–1625

Haxby JV, Horwitz B, Ungerleider LG, Maisog JM, Pietrini P, Grady CL (1994a) The functional organization of human extrastriate cortex: a PET rCBF study of selective attention to faces and locations. J Neurosci 14: 6336–6353.

Haxby JV, Ungerleider LG, Horwitz B, Maisog JM, Grady CL (1994b) Neural systems for encoding and retrieving new long-term visual memories: a PET-rCBF study. Invest Ophthalmol Vis Sci 35: 1813

Haxby JV, Ungerleider LG, Horwitz B, Rapoport SI, Grady CL (1995) Hemispheric differences in neural systems for face working memory: a PET-rCBF study. Human Brain Map 3: 68–82

Horwitz B, Grady CL, Haxby JV, Ungerleider LG, Schapiro MB, Mishkin M, Rapoport SI (1992) Functional associations among human posterior extratriate brain regions during object and spatial vision. J Cog Neurosci 4: 311–322

Jonides J, Smith EE, Koeppe RA, Awh E, Minoshima S, Mintun MA (1993) Spatial working memory in humans as revealed by PET. Nature 363: 623–625

McIntosh AR, Grady CL, Ungerleider LG, Haxby JV, Rapoport SI, Horwitz B (1994) Network analysis of cortical visual pathways. J Neurosci 14: 655–656

Nobre AC, Allison T. McCarthy G (1994) Word recognition in the human inferior temporal lobe. Nature 372: 260–263

Scoville WB, Milner B (1957) Loss of recent memory after bilateral hippocampal lesions. J Neurol Neurosurg Psychiat 20: 11–21

Sereno MI, Dale AM, Reppas JB, Kwong KK, Belliveau JW, Brady TJ, Rosen BR, Tootell RB (1995) Borders of multiple visual areas in humans revealed by functional magnetic resonance imaging. Science 268: 889–893

Squire LR (1992) Memory and the hippocampus: A synthesis from findings with rats, monkeys and humans. Psychol Rev 99: 195–231

Tootell RBH, Reppas JB, Kwong KK, Malach R, Born RT, Brady TJ, Rosen BR, Belliveau JW (1995) Functional analysis of human MT and related visual cortical areas using magnetic resonance imaging. J Neurosci 15: 3215–3230

Tulving E, Kapur S, Craik FIM, Moscovitch M, Houle S (1994) Hemispheric encoding/retrieval asymmetry in episodic memory: Positron emission tomography findings. Proc Natl Acad Sci USA 91: 2016–2020

Ungerleider LG, Haxby JV (1994) 'What' and 'where' in the human brain. Curr Opin Neurobiol 4: 157–165

Ungerleider LG, Mishkin M (1982) Two cortical visual systems. In: Ingle DJ, Goodale MA, Mansfield RJW (eds) Analysis of visual behavior. MIT Press, Cambridge, pp 549–586

Watson JD, Meyers R, Frackowiak RSJ, Hajnal JV, Woods RP, Mazziotta JC, Shipp S, Zeki S (1993) Area V5 of the human brain: evidence from a combined study using positron emission tomography and magnetic resonance imaging. Cereb Cortex 3: 79–94

Wilson FA, Scalaidhe SP, Goldman-Rakic PS (1993) Dissociation of object and spatial processing domains in primate prefrontal cortex. Science 260: 1955–1958

Zeki S, Watson JPG, Lueck CJ, Friston K, Kennard C, Frackowiak RSJ (1991) A direct demonstration of functional specialization in human visual cortex. J Neurosci 11: 641–649

Language and the Alzheimer Brain

A. R. Damasio and *H. Damasio*

Impairments of language are a common finding in patients with Alzheimer's disease and it is thus important to gain a modern perspective on the functional and anatomo-pathological meaning of those impairments. It is especially important to incorporate in our picture of Alzheimer's disease the new data on the neural basis of language, which reveal that specific damage to left temporal and frontal sites causes defects in word retrieval, by far the most common language impairment in Alzheimer's disease. The following text is a review of recent progress.

Language is the ability to use words (or signs from a sign language) and combine them in sentences so that the concepts we experience in our minds can be transmitted to other individuals; or so that, in reserve, the concepts in the minds of others can be rendered in words and sentences and turned into concepts in our own minds.

Language exists in two distinct aspects: as an artifact in the world external to us, a collection of symbols or utterances in rule-governed combinations; and as the embodiment of those symbols and rules in our brain structure. Noam Chomsky, the most influential linguist of our time, called the former E-language (E for external) and the latter I-Language (I for internal; Chomsky 1985). We believe words (or signs) and the principles that govern their structure and combination (grammatical operations) are represented in the brain using precisely the same neural machinery of representation that is used for another entity, or for rules of admissible relationship among those entities in spacetime. If the brain contains representations of external nonlanguage entities, events, and their relations, *and* also contains representations of language, then the main issue in the neuroscience of language is understanding the neural basis for 1) those two sets of representations and 2) their interrelation.

By now the reader will have guessed our answer to the question about which came first, language or concepts? And the answer is that in the beginning there were no words, just actions and just concepts. Language seems to have appeared in evolution only after the brains of our species, and of many other species before ours, were adept at generating and categorizing actions to interact with their surroundings and also adept at representing and categorizing things, events, and

* Department of Neurology (Division of Behavioral Neurology and Cognitive Neuroscience) University of Iowa College of Medicine Iowa City, Iowa

B. T. Hyman / C. Duyckaerts / Y. Christen (Eds.)
Connections, Cognition, and Alzheimer's Disease
© Springer-Verlag Berlin Heidelberg 1997

relationships in that same surrounding. Concepts of varied order, from simple and concrete to complex and abstract, must have come first. In the beginning there was knowledge but no language.

The idea that concepts preceded language applies not only to the history of evolution (i.e., to phylogeny) but also to our own development from birth to adulthood (i.e., to ontogeny). Infants' brains are busy representing and evoking concepts, and generating myriad actions, long before they utter, their first well-selected word and even longer before they truly use language, i.e., form a sentence. New evidence on previously normal persons who develop language *or* concept disorders as a result of brain lesions sheds some light on the relation between concepts and language (Damasio and Damasio 1989). After certain lesions, concepts remain intact even though language fails. Other lesions cause concepts to fail but language can remain intact, including sometimes the very words that denote the compromised concepts. This does not mean that the maturation of all language processes is necessarily subsidiary to the maturation of conceptual processes. There is evidence that children with deficient conceptual systems have nonetheless acquired the grammar. The impression left by these findings is that the neural systems on which syntactic operations depend can learn and generalize their own domain of knowledge even when many of the neural systems on which concepts depend are undeveloped. This is about as we can get to evidence for a relatively autonomous language device.

What advantage did language confer on our species? Efficiency of communication is the most obvious answer. Humans have several means of expressing concepts other than language. They can use facial expressions, hand gestures, or body movement; they can vocalize wordless calls; they can play with objects, or draw. However, the efficacy of communication based on those means of expression is poor. Abstract concepts and complex ideas cannot be communicated properly, and communication of lengthy messages takes an inordinate amount of time and remains vague. As an exercise, try to explain the rise and fall of the communist republics, without using a single word, and see how far you get. But language offers other advantages. For instance, it serves as a supplementary means of categorization, and it reduces the complexity of conceptual structure to a manageable scale. Consider how the word "screwdriver" stands for a large collection of evocable representations of such an instrument, e.g., the visual descriptions of its operation and outcome; specific instances of its use; the feel of the tool or the hand movement that pertains to it. Or consider how language helps you narrate internally, in sentences, a large number of concepts which do not even need to be fully evoked into consciousness because language swiftly stands for them and holds them in continuity for our reasoning. Think also about the immense number and variety of conceptual representations that are denoted by a word such as "democracy". There is little doubt that the reductory and classificatory roles of language, its cognitive economies, open the way for establishing and operating on concepts at ever more complex and abstract levels. Concepts came first, but not *all* concepts.

Which language components can we hope to find represented in the human brain? Considering an auditory-based language, such as English, they include: 1) the sound system, known as phonology, which includes individual speech sounds, phonemes, and prosody, the vocal intonation which provides words and sentences with meanings beyond their basic dictionary definition; 2) the smallest meaningful units of a word, known as morphemes, and the way they can be put together to form words, known as morphology; 3) the rules of combination of words to form sentences, which is syntax (or grammar, in its popular usage); and 4) the lexicon, the dictionary of words available in a language. Some linguists would also include a discourse component that accounts for the linking of sentences such that they constitute a narrative. (The same would apply to sign languages, such as American Sign Language, by substituting a visuomotor system where the sound system is (Klima and Bellugi 1978). In sign language, morphology creates a "sign" rather than a word, although it is not inappropriate to refer to signs as words and to their combinations as sentences.)

From this listing (culled out of Victoria Fromkin's standard textbook; Fromkin and Rodman 1993) it is clear that language is not simply a collection of words. This is an important distinction since language is all too often mistaken for the mere capacity to denote an object with a name. Language does revolve around words but it also encompasses components above and below the word level. For auditory-bases language, the subcomponents include phonetic features that combine to form *phonemes*, the individual sound units whose concatenation in a particular order produces morphemes, and *morphemes*, whose combination creates a *word*. Take as an example the word *bakers*. It comprises three morphemes: a free or base morpheme, *bake*, and two bound morphemes, *-er*, a derivational morpheme, and *-s*, an inflectional morpheme. The base morpheme "bake" is the one that refers to the concept of baking. The bound morpheme "er" tells us that the word is not just about the baking action but specifically about a person who performs that action. The other bound morpheme "s" indicates that there is more than one such person.

The next component is *syntax*. It pertains to the admissible combinations of words in sentences and to the allowable ways of ordering words within sentences. Since it is the job of language to represent not only *materials* (things, substances) and *situations* (states, activities, events), but also *relationships among materials and situations in spacetime*, we must have a way of combining words so that those relationships can be described. Syntax does just that.

The next component, the lexicon, is the collection of all words in a given language. The lexical entry for each word includes all information that has morphological or syntactic ramifications. Unlike an ordinary dictionary, a linguistic lexicon does *not* include conceptual meanings.

The words and sentences that can be formed with these components then connect with concepts, i.e., nonlinguistic meanings. The sum total of the meanings that correspond to all lexical items, and to the infinite number of sentences you can create or hear with the lexical items, is called *semantics*. The conceptual meanings largely stand, in our view, *outside* language. They are what language bridges to and from, the universe of knowledge that can be rendered in words and sentences (when we want to translate our thoughts) and that can be evoked internally (when we hear a sentence or read it).

How are things organized in the brain such that we can turn concepts into language and also be able to render the words and sentences of others into concepts in our minds? We believe there are three essential sectors to consider, as follows:

1. A restricted number of neural systems, usually and predominantly located in the left cerebral hemisphere, represents phonemes, phoneme combinations, and rules for word-form combinations (syntactic structures). When they are stimulated from within the brain, these systems assemble word-forms and generate sentences, in both acoustical and motor forms. When they are stimulated externally, these systems' output is toward mediational systems.
2. A far more extensive number of neural systems in both right and left cerebral hemispheres, represents nonlanguage interactions between our body and its surroundings as mediated by varied sensory and motor systems (those interactions describe concepts, entities and events outside our bodies, correlated body states, and their myriad relationships in spacetime). Those nonlanguage representations also include the categorization of all those concepts across many levels of complexity, and form the basis for abstraction and metaphor.
3. A third collection of neural systems, located largely in the left cerebral hemisphere, mediates between the two sets of systems described above. It evokes word-forms in connection with concepts, and vice versa; it turns a set of relationships among concepts into sentences and vice versa.

It is encouraging to discover that a tripartite organization such as we propose here has also been hypothesized from a psycholinguistic perspective, in which reference to brain facts plays no direct role. William Levelt has suggested that word-forms and sentences would be generated from conceptual processing through the agency of a third component that he called *lemma*. Merrill Garret holds a similar view.

Let us proceed by discussing additional evidence for this tripartite arrangement in the human brain. The evidence for systems that generate word-forms and sentences dates back to the discoveries of Paul Broca and Karl Wernicke over a century and a half ago, and their pioneering findings have been largely validated in recent years. The evidence for the neural basis of conceptual structure is a recent topic of research. The evidence for third-party mediational systems is the newest and most exciting.

Neural Systems for the Implementation of Word-Forms and Sentences

The fact that in most humans language structures are largely in the left hemisphere rather than the right constitutes a brain disposition known as cerebral dominance for language. This disposition is strongly correlated with right-handedness (it is unlikely that more than one percent of right-handers ever violate it) and is even found in about two-thirds of left-handers (Damasio 1992). The study of aphasic patients with different language backgrounds hallmarks the strength of these facts. In fact, a notable development in the neurobiology of language is the discovery of aphasia in relation to a nonauditory-based language, American Sign Language (ASL).

The neural systems most directly involved in word-form implementation and sentence formation are mostly confined to the left perisylvian region and subjacent gray nuclei. The evidence comes from lesion studies in aphasic patients, and electrophysiologic recording and stimulation of exposed cerebral cortex during surgery for epilepsy. The posterior perisylvian sector includes areas 41, 42, 22 (Wernicke's area), 40, and part of areas 37 and 39. Damage in this sector disrupts the assembly of phonemes and the selection of entire word-forms. As a consequence, many words are improperly formed (e.g., "loliphant" for elephant), or substituted by a pronoun or a word at a more general taxonomic level (the supraordinate "people" for woman), or semantically related ("headman" for president), or not produced at all. However, the words are produced at normal rates, with normal prosody, and the sentences' syntactical frames are undisturbed (even if there may be errors in the use of functor words such as pronouns and conjunctions). Damage here also impairs processing of speech sound input, and as a consequence, auditory comprehension of words and sentences is defective. Auditory comprehension fails, not because the posterior perisylvian sector is a center to store "meanings" of words, as has been traditionally stated, but because the normal acoustical analyses of the word-forms one hears are aborted at an early stage.

We believe that the systems in this posterior sector normally hold auditory and kinesthetic records of phonemes and phoneme sequences. The auditory and kinesthetic records, based on areas 22 and 40, respectively, can be cross-activated by reciprocal neural projections. In turn both are interconnected with motor and premotor cortics and both project to the left basal ganglia, which are part of a subcortical circuit that includes anterolateral nuclei in left thalamus and projects back to motor cortex. This dual motor route is especially important. It means that the motor implementation of speech sounds can utilize either a cortical or a subcortical circuit, or both. The subcortical would be the fast, "habit-learning" route. For instance, when a child learns a word-form such as "yellow," all of these systems would be simultaneously active and their activities correlated with the left medial occipitotemporal systems in which color concepts and concept/language mediation take place. In time, we suspect that the posterior perisylvian sector need not be strongly activated to produce a word-form because the mediation unit would use a direct route to basal ganglia. Subsequent learning of the word-form for the color yellow in another language would again require participation of the perisylvian region so that auditory, kinesthetic, and motor correspondences of phonemes could be established. (It is likely that both cortical so-called "associative" and subcortical "habit" systems operate in parallel during language processing, one or the other being preferred depending on the history of language acquisition and the nature of the item). Shifting the implementation of a color name, for instance, from an elaborate and slow cortical network to a faster and automated subcortical circuit is an example of the former. An example of the latter is the assignment of irregular verbs *(take, took, taken)* to associative learning, and regular verbs (*-ed* verbs) to habit learning.

The anterior perisylvian sector includes areas 44 and 45 (Broca's area) and nearby areas 47 and 6, 46, 8, 9, and 10. As was the case with the posterior perisylvian sector, the left basal ganglia are part and parcel of this sector, and there are reasons to believe that the entire sector is strongly associated with the cerebellum, although the role of the latter remains to be elucidated. Patients with damage in the anterior sector produce melodically flat, sparse, and agrammatic speech. Functor words are usually missing; grammatical order is often compromised; there may be long pauses between words. Nouns come easier to patients with these lesions than to verbs (Damasio and Damasio 1989; Damasio 1992). Under certain circumstances patients with damage in the anterior perisylvian sector are also compromised in their ability to grasp the meaning of syntactic structures.

Neural System for Concept Representation

We believe that concepts are represented in coded form and that there are no permanently held "pictorial" representations of objects or persons as was traditionally thought. Instead, the brain is likely to keep the equivalent of a record of parameters of neural activity in the sensory and motor cortices which are engaged during a perceiver's interaction with a given object. We have proposed that the brain inscribes signals of our sensory and motor interactions with the universe in whichever neural systems are appropriate to define the encounters of body and object. The records are fragmentary and coded. They are patterns of synaptic connections on the basis of which: (the separate sets of activity which defined an object or event can be momentarily reactivated in different parts of the brain, at roughly the same time; and 2) related records can be activated.

For example, in our interaction with a coffee cup, the brain will be active in visual cortices in response to the color of the cup and of its contents, and to the cup's shape and spatial position. But the somatosensory cortices, which are concerned with touch, temperature, and body state, will also be active in response to the shape the hand assumes as it holds the cup, to the movement of hand and arm as they bring the cup to the lips, to the warmth of the coffee, and to the body change we call pleasure when we drink the stuff. Note that the brain is not merely representing aspects of external reality but is making records of how the body explores and is modified by reality. The neural processes that describe the interaction between the individual and the object happen in a rapid sequence of micro-percepts and micro-actions, which are almost simultaneous as far as the time scale of consciousness is concerned. These neural processes thus occur in separate functional regions, e.g., visual, somatosensory, motor, and there are further functional subdivisions within each of them, e.g., the visual component is segregated within smaller systems partially specialized for color, shape and movement. There are similar segregations for touch and movement. Where can the records that bind all these fragmented activities be held? We believe the binding records are embodied in neuron ensembles within many "convergence" regions. These are neural sites where the axons of feedforward projecting neurons can converge and be conjoined with reciprocally diverging feedback projecting neurons. Subsequently, when re-activation within convergence zones causes the feedback projections to fire simultaneously, anatomically separate and widely distributed neuron ensembles can also fire synchronously and in so doing reconstruct previous patterns of multiregional activity.

Every aspect to the interaction described above can be categorized, which means that anything that belongs together can henceforth be reactivated together, e.g., shapes, colors, trajectories in spacetime, pertinent body movements and body reactions. The categorization is denoted by yet another coded record, in yet another interconnected convergence zone. The essential properties of the entities and processes in any interaction can thus be represented separately but interwovenly, and symbolic representations such as metaphor can emerge from this architecture. Examples of what is represented in this manner include: the fact that the cup has *dimensions* and has a *boundary;* that it is *made of something* and it has *parts;* that if it is divided it no longer is a cup, unlike a *substance* such as water, whose parts are not visible but which will remain water no matter how many times we divide a certain amount of it; the fact that the cup moved from an *inception point,* with a *direction in space,* towards a *conclusion point;* the fact that the conclusion of the movement produced a specific *outcome.*

Concept/Language Mediation Systems

The concept/language mediation systems that we envision work in both directions and operate not only to bring about correct lexical selection but also to pace the generation of syntactical structure. They do not implement word-form or syntactical structure themselves, which is the job of word-form/sentence implementation units, but they drive those structures and can be driven by them following the processing of language input. The evidence for this neural brokerage is beginning to emerge from the study of neurological patients. Here we will only discuss evidence for lexical mediation systems and leave aside the issue of structural mediation. Consider the example of patients AN and LO. Both patients have normal retrieval of concepts. When shown pictures of entities or substances from virtually any conceptual category (human faces, body parts, animals and botanical specimens, vehicles and buildings, tools and utensils), AN and LO know unequivocally what they are looking at. For each item, they can define whatever functions, habits, habitats, and value best apply. They perform no differently than normal controls. If they are given sounds corresponding to those entities and substances (whenever a sound happens to be associated with them), AN and LO are equally able to know what entity or substance is in question. The same happens even when they are blindfolded and asked to recognize an object placed in their hands. However, in spite of their remarkable show of knowledge, they have difficulty in retrieving the names for many of the objects they know so well. On the average they come up with less than 50 % of the names they ought to retrieve. In short, these patients have a compromised ability to access the word-form which denotes the objects. Their conceptual knowledge may be fine, but access to the lexicon is not.

But there are more surprising dissociations to come. First, the deficit in word-form retrieval is not of the same magnitude for the different conceptual categories. As an example, nouns that denote tools and utensils are easier to retrieve than nouns for animals, fruits and vegetables.

Perhaps the greatest surprise is to discover that these patients have no difficulty in producing words for actions, or that they are equally normal at the production of functors (words such as prepositions, conjunctions, pronouns), or that their sentences are perfectly formed and grammatical. As they speak or write, they produce a narrative in which, instead of the missing noun, they will substitute words like "thing" or "stuff", or pronouns such as "it" or "she" or "they". But the verbs that animate the arguments of those sentences are properly selected and produced, and properly marked as far as tense (present, past, future) and person. In other experiments whose results have been repeatedly confirmed, these patients perform precisely as matched controls do in tasks requiring them to generate words for actions. The relation of words in a sentence and the order those words should occupy in grammatical English are impeccable. So are, by the way, the morphology of the words (their appropriate phonemic structure), and their prosody (the intonation of the individual words and the entire sentence). Patients AN and LO suffer from a selective form of anomia.

Where are the lesions in these patients? In both cases the lesions destroyed the middle and anterior sectors of the left temporal lobe and thus spared the cortices on which word-form and sentence implementation depend. We propose that these regions contain lexical mediation systems of "proper nouns" and "common nouns" denoting entities of particular class (e.g., visually ambiguous, nonmanipulable entities such as most animals are).

We have now seen many patients with a defect in word retrieval for unique persons but *without* a disturbance of word retrieval for non-unique items. Their lesions are restricted to the left temporal pole and medial temporal surface, and spare the lateral and inferior temporal lobe. This is further evidence for the regionalization of lexical mediation systems (Damasio et al. 1990).

It goes without saying that if patients AN and LO can retrieve words for actions and functor words normally, then the regions required for access to such words cannot be in the left temporal region. They must be accessed from other regions in the network, and the preliminary evidence points to frontal and parietal sites. This finding is bases on aphasia studies that reveal how patients with left frontal damage have far more trouble with verb than with noun retrieval. Indirect evidence comes from PET studies (Petersen et al. 1988) in which normal subjects who were asked to generate a verb corresponding to the picture of an object (e.g., given *apple*, produce *eat*) activated a region of lateral and inferior dorsal frontal cortex. Curiously, damage to those regions not only compromises access to verbs and functors but also disturbs the grammatical structure of sentences. It makes entire sense that the mediational systems for syntax would overlap with those concerned with verbs and functors, which constitute the core of syntactical structure.

All of the above results have been extensively replicated in both patients with lesions and normal individuals who participated in PET experiments (Damasio et al. 1996). They provide a clear indication that the pronounced naming defects seen in patients with Alzheimer's disease are probably caused by damage in higher-order association cortices outside the hippocampal system and also outside the classic language areas.

Acknowledgments

The work described here was supported in part by the Kiwanis International and the Spastics Paralysis Research Foundation.

References

Chomsky N (1985) Knowledge of language. Praeger, New York

Klima E, Bellugi U (1978) The signs of language. Cambridge, Massachusetts, Harvard University Press

Fromkin V, Rodman R (1997) Introduction to language. 6th Edition. Harcourt, Brace, Jovanovich, New York

Damasio H, Damasio AR (1989) Lesion analysis in neuropsychology. Oxford University Press, New York

Damasio AR, Damasio H, Tranel D, Brandt JP (1990) Neural regionalization of knowledge access: Preliminary evidence. Symposia on quantitative biology, Vol. 55. Cold Spring Harbor Laboratory Press, pp 1039–1047.

Damasio AR (1992) Aphasia. New Engl J Med 326: 531–539

Damasio AR, Tranel D (1993) Nouns and verbs are retrieved with differently distributed neural systems. Proc Nat Acad Sci USA 90: 4947–4960

Damasio H, Grabowski TH, Tranel D, Hichwa R, Damasio AR (1996) A neural basis for lexical retrieval. Nature 380: 499–505

Petersen SE, Fox PT, Posner MI, Mintun M, Raichle ME (1988) Positron emission tomographic studies of the cortical anatomy of single-word processing. Nature 331: 585–589

Subject Index

Printing: Saladruck, Berlin
Binding: Buchbinderei Lüderitz & Bauer, Berlin